Series Editor
John M. Walker
School of Life Sciences
University of Hertfordshire
Hatfield, Hertfordshire, AL10 9AB, UK

For further volumes:
http://www.springer.com/series/7651

Polyadenylation in Plants

Methods and Protocols

Edited by

Arthur G. Hunt

Department of Plant and Soil Sciences, University of Kentucky, Lexington, KY, USA

Qingshun Quinn Li

Department of Biology, Miami University, Oxford, OH, USA
Key Laboratory of the Ministry of Education for Coastal and Wetland Ecosystems, College of the Environment and Ecology, Xiamen University, Xiamen, Fujian, China

 Humana Press

Editors
Arthur G. Hunt
Department of Plant and Soil Sciences
University of Kentucky
Lexington, KY, USA

Qingshun Quinn Li
Department of Biology
Miami University
Oxford, OH, USA

Key Laboratory of the Ministry of Education for
Coastal and Wetland Ecosystems
College of the Environment and Ecology
Xiamen University
Xiamen, Fujian, China

ISSN 1064-3745   ISSN 1940-6029 (electronic)
ISBN 978-1-4939-4668-6   ISBN 978-1-4939-2175-1 (eBook)
DOI 10.1007/978-1-4939-2175-1
Springer New York Heidelberg Dordrecht London

Preface

Messenger RNA polyadenylation is an important aspect of gene expression in eukaryotes. The process itself is physically and temporally coupled with several other steps in gene expression, ranging from the initiation of transcription to the export of the mature mRNA to the cytoplasm. In the cytoplasm, the poly(A) tail constitutes an essential cis-element that enables the translation of the mRNA, in cooperation with the 5′-cap.

mRNA polyadenylation is an RNA processing event whereby a precursor mRNA is cleaved and a poly(A) tract added to the 3′ end of the cleaved RNA. The position along the precursor mRNA where processing occurs contributes to the overall nature of the mature mRNA. Typically, the poly(A) site is situated several tens or hundreds of nucleotides 3′ of the translation termination codon that is specific within the protein-coding portion of the mRNA. However, for a majority of genes in eukaryotes, the primary transcript may be processed and polyadenylated at more than one position; the result of this multiplicity is a set of mRNA isoforms, each of which has a different 3′-UTR as well as other mRNA features. This multiplicity of mRNA isoforms gives rise to the possibility that gene expression may be controlled in part by alterations in the 3′ end profiles of expressed genes. This possibility has been confirmed in many systems; perhaps most prominent along these lines is the realization that alternative poly(A) site choice is a key determinant of the levels of expression of growth-promoting oncogenes in cancer cells.

Recent years have seen a renewed interest in the process of mRNA polyadenylation in plants. There have been a growing number of reports that link polyadenylation with several important molecular and physiological processes. Moreover, with the advent of genome-scale studies extending back more than 10 years and spanning the "ages" of large-scale EST studies, to microarray-based experiments, and more recently to high-throughput sequencing, it has become apparent that alternative poly(A) site choice may be a common feature of gene expression in plants, and that alterations in 3′ end profiles may contribute to the regulation of gene expression in several ways. For these reasons, the field has grown in terms of scope and methodology, and now encompasses a broad spectrum of experimental approaches.

In this volume, the breadth of these studies is encapsulated so that interested readers may appreciate and adapt the tools and outcomes, and also gain a better feel for the current state of the field. The chapters in this volume are grouped into three parts. The first part represents a relatively recent development in the field, namely the use of bioinformatics and computational tools to study polyadenylation. These tools include methods for analyzing and predicting plant polyadenylation signals (Chapters 1 and 2) and for retrieving poly(A) sites from public sequence databases (Chapter 3). Chapter 4 describes computational approaches for analyzing large-scale sequencing data specifically designed for poly(A)-tag sequencing (PAT-seq), while Chapter 5 describes a computational assay that uses PAT-seq to measure poly(A) site choice on a genome-wide scale. Chapter 6 presents a comprehensive approach for studying alternative mRNA processing, including polyadenylation, using data derived from tiling microarrays.

The second part brings into focus the numerous molecular, biochemical, and cellular methods that have been used to characterize polyadenylation in plants. Described in this part are techniques used to characterize polyadenylation complexes and their activities in vitro (Chapters 7 and 8) and basic operation to analyze related proteins (Chapter 9), to study novel aspects of polyadenylation factor structure and function using chemical modification and mass spectroscopy (Chapter 10), for transient expression to study polyadenylation proteins and polyadenylation signals (Chapter 11), and for the determination of mRNA 3′ ends using a modification of the so-called 3′-RACE assay (Chapter 12). The third part consists of a compendium of methods used to characterize polyadenylation sites on a genome-wide scale. A high-throughput approach to identifying protein interacting partners, and further to identifying small sequence motifs that may be associated with such interactions, is described in Chapter 13. Chapters 14–16 describe three methods for the generation of sequence tags that query the mRNA-poly(A) junction on a genome-wide scale; these sequence tags are suitable for the so-called next generation sequencing technology. Along with poly(A) tag analyses, we also need to perform RNA-seq to get sequence information beyond poly(A) sites. Thus, a simple, inexpensive protocol for constructing RNA-seq libraries is also included as Chapter 17 . Chapter 18 was used to study the relationship between transcription and polyadenylation through RNA polymerase II activities.

It is our hope that this volume will bring together, in one place, the range of methods that are used to study polyadenylation in plants (or other organisms), and in so doing will provide readers convenient ways to employ these tools to advance the understanding of the role of mRNA polyadenylation in plant gene expression.

Lexington, KY, USA
Oxford, OH, USA

Arthur G. Hunt
Qingshun Quinn Li

Contents

PART III   GENOME-SCALE STUDY OF POLYADENYLATION IN PLANTS

Contributors

Balasubrahmanyam Addepalli • *Rieveschl Laboratories for Mass Spectrometry, Department of Chemistry, University of Cincinnati, Cincinnati, OH, USA*

Stephen A. Bell • *Department of Pharmaceutical Sciences, University of Kentucky, Lexington, KY, USA*

Jingyi Cao • *Department of Biology, Miami University, Oxford, OH, USA*

Laura de Lorenzo • *Department of Plant and Soil Sciences, University of Kentucky, Lexington, KY, USA*

Min Dong • *Department of Automation, Xiamen University, Xiamen, Fujian, China; Department of Biology Miami University, Oxford, OH, USA*

Arthur G. Hunt • *Department of Plant and Soil Sciences, University of Kentucky, Lexington, KY, USA*

Guoli Ji • *Department of Automation, Xiamen University, Xiamen, Fujian, China*

Qingshun Quinn Li • *Department of Biology, Miami University, Oxford, OH, USA; Key Laboratory of the Ministry of Education for Coastal and Wetland Ecosystems, College of the Environment and Ecology, Xiamen University, Xiamen, Fujian, China; Rice Research Institute, Fujian Academy of Agricultural Sciences, Fuzhou, Fujian, China*

Chun Liang • *Department of Biology, Miami University, Oxford, OH, USA; Department of Computer Science and Software Engineering, Miami University, Oxford, OH, USA*

Man Liu • *Department of Biology, Miami University, Oxford, OH, USA*

Liuyin Ma • *Key Laboratory of the Ministry of Education for Coastal and Wetland Ecosystems, College of the Environment and Ecology, Xiamen University, Xiamen, Fujian, China; Department of Plant and Soil Sciences University of Kentucky, Lexington, KY, USA*

Pratap Kumar Pati • *Department of Biotechnology, Guru Nanak Dev University, Amritsar, India; Department of Horticulture University of Kentucky, Lexington, KY, USA*

Suryadevara Rao • *Department of Plant Pathology, University of Kentucky, Lexington, KY, USA*

Patrick E. Thomas • *Department of Plant and Soil Sciences, University of Kentucky, Lexington, KY, USA; Franklin-Simpson High School, Franklin, KY, USA*

Carol Von Lanken • *Department of Plant and Soil Sciences, University of Kentucky, Lexington, KY, USA*

Xiaohui Wu • *Department of Automation, Xiamen University, Xiamen, Fujian, China*

Denghui Xing • *Department of Biology, Miami University, Oxford, OH, USA; Department of Biology Colorado State University, Fort Collins, CO, USA*

Xinfu Ye • *Fruit Research Institute, Fujian Academy of Agricultural Sciences, Fuzhou, Fujian, China*

Hongwei Zhao • *College of Plant Protection, Nanjing Agricultural University, Nanjing, Jiangsu, China*

Part I

Bioinformatics Studies of Plant Polyadenylation

Chapter 1

Computational Analysis of Plant Polyadenylation Signals

Xiaohui Wu, Guoli Ji, and Qingshun Quinn Li

Abstract

Messenger RNA polyadenylation in eukaryotes marks the end of a transcript, and the process is associated with transcription termination. Increasing evidence reveals the potential of gene expression regulation through alternative polyadenylation. The site of poly(A) addition is defined by poly(A) signals reside in the transcribed pre-mRNA. To gain further insight into poly(A) signals and their functions in defining alternative polyadenylation sites that lie within different genomic regions, SignalSleuth2 was developed to extract and analyze cis-elements from a set of data with known poly(A) sites. After obtaining the sequences surrounding the poly(A) sites, exhaustive search of short sequence motifs in specified range of nucleotide sequences are performed, variable motif sizes and rank the detected motifs based on their occurrence frequencies are tallied. It also has new functions including Position-Specific Scoring Matrix (PSSM) scores calculation and multiple scanning modes. This program is powerful in revealing underline sequence motifs surrounding any target regions in a given dataset.

Key words Polyadenylation signal, Pattern recognition, Alternative polyadenylation, Cis-elements

---

1 Introduction

The polyadenylation of messenger RNA in eukaryotes is an essential step in gene expression. Polyadenylation is guided by cis-acting elements surrounding the poly(A) site [1], collectively known as the poly(A) signals. Besides poly(A) signals, the 3′-untranslated regions (3′-UTRs) containing cis-acting elements that may interact with RNA binding proteins and small noncoding RNAs, thereby affecting the function of RNA, such as mRNA stability, exportation, localization, and translatability [2–7]. However, the selection of an alternative poly(A) sites may change some of these cis-acting elements in the mature mRNA. Thus, finding where poly(A) signals are located would help us to understand how alternative polyadenylation is regulated.

It is understood that a poly(A) signal is in the vicinity of a poly(A) site [8]. To study the signals for mRNA poly(A) tailing,

Arthur G. Hunt and Qingshun Quinn Li (eds.), *Polyadenylation in Plants: Methods and Protocols*, Methods in Molecular Biology, vol. 1255, DOI 10.1007/978-1-4939-2175-1_1, © Springer Science+Business Media New York 2015

it is necessary to analyze the nucleotide patterns in the poly(A) site-related signal regions and select useful features from a large number of nucleotide sequences. Currently there are many methods for poly(A) signal or poly(A) site recognition in human [1, 8–11]. Legendre and Gautheret [9] developed a program called Erpin which used 2-g position-specific nucleotide acid patterns to characterize the sequences around candidate poly(A) signals. Liu et al. [10] selected *k*-grams by an entropy-based algorithm and utilized support vector machine SVM to classify poly(A) sites. Cheng et al. [12] used position specific scoring matrix (PSSM) to characterize patterns and also used SVM predict poly(A) sites. Akhtar et al. [11] classified poly(A) sites into three classes and developed POLYAR program to do the prediction.

In this chapter, we present a workflow to study the poly(A) signals in plants. Firstly, the relative base compositions of the sequences surrounding these sites are studied. A new version of SignalSleuth [14], SignalSleuth2, was developed to perform exhaustively search of short sequence motifs in specified range of nucleotide sequences with variable motif sizes (generally 3–8 nt in length) and rank the detected motifs based on their occurrence frequencies. In addition, SignalSleuth2 has new functions including (PSSM) scores calculation and multiple scanning modes. Occasionally, a target motif may appear multiple times within a given region of a sequence. Sometimes, such multiple occurrences might be overlapped, resulting in over-representation of a specific motif. SignalSleuth2 provides a distance parameter (*-gap*) to prevent over-counting of the overlapping motifs. While the program was tested in the analysis of plant poly(A) signals, it can be implemented in any cis-element analysis.

2   Materials

2.1   Equipment

Hardware: a computer running Linux or MS Windows.

Please make sure that Perl (http://www.perl.org/) is installed in your system.

2.2   Data

8 K dataset or other data sets can be downloaded from the following links http://www.users.miamioh.edu/liq/links.html or http://www.polyA.org.

2.3   Setting Up

Two Perl scripts: PAS_kpssm.pl and SignalSleuth2.pl. can be downloaded from the above links. These scripts can be in any directory.

3   Methods

3.1 Computational Analysis of Plant Polyadenylation Signals

SignalSleuth2 was developed for an exhaustive search of short sequence motifs. Three *k*-gram scanning modes are used to filter out useless *k*-grams (Fig. 1), including gap scanning mode, overlapping scanning mode, and once scanning mode. In "gap scanning mode," SignalSleuth2 provides a distance parameter (-gap) to prevent over-counting of the overlapping motifs. For example, ATATAT will be counted only once in the sequence ATATATAT if -gap is set to be 5. For another example, if AATAAA motif is searched, -gap = 5 will avoid over counting overlapping motifs if they exist. In "overlapping scanning mode" (namely, -gap = 0), the frequencies of overlapping signals can be obtained. In "once scanning mode," only the motif that is the closest to a poly(A) site is chosen if there are more than one occurrence of non-overlapped motif in a given sequence region.

For each scanning mode, SignalSleuth2 provides PSSM results simultaneously to detect whether a motif is representative in the studied region for a given sequence dataset. The workflow to calculate the PSSM score of a motif is shown in Fig. 2. First, a given region of the sequence is scanned for the presence of the motif and the score of the motif is then calculated using the PSSM generated from that motif. The score of each position of the given region is the average of all positive scores of all sequences in the given dataset. Finally, the scores are smoothed by a sliding window with a length of 3 nucleotides. To calculate the PSSM score, each motif is used to generate a PSSM with dimension $4 \times L$, where L is the length of the motif. For a given sub-sequence with the length equal to the column number of the PSSM, its score is the sum of individual scores at all nucleotide positions. Higher score indicates the higher likelihood of the presence of a motif.

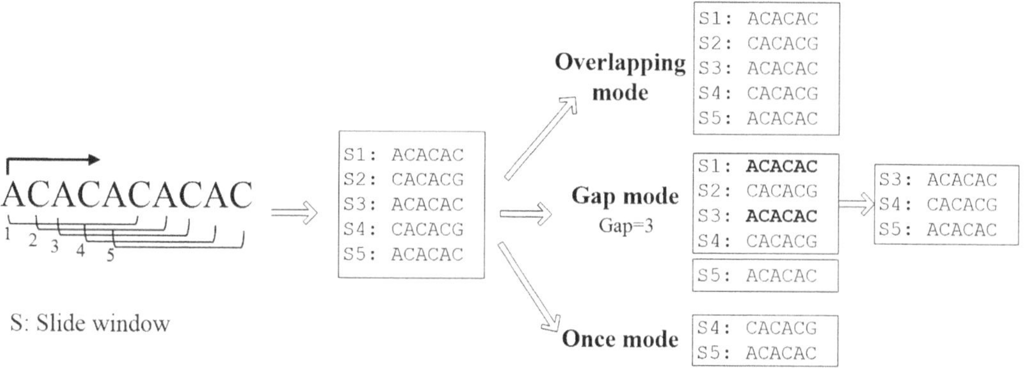

Fig. 1 Three *k*-gram scanning modes

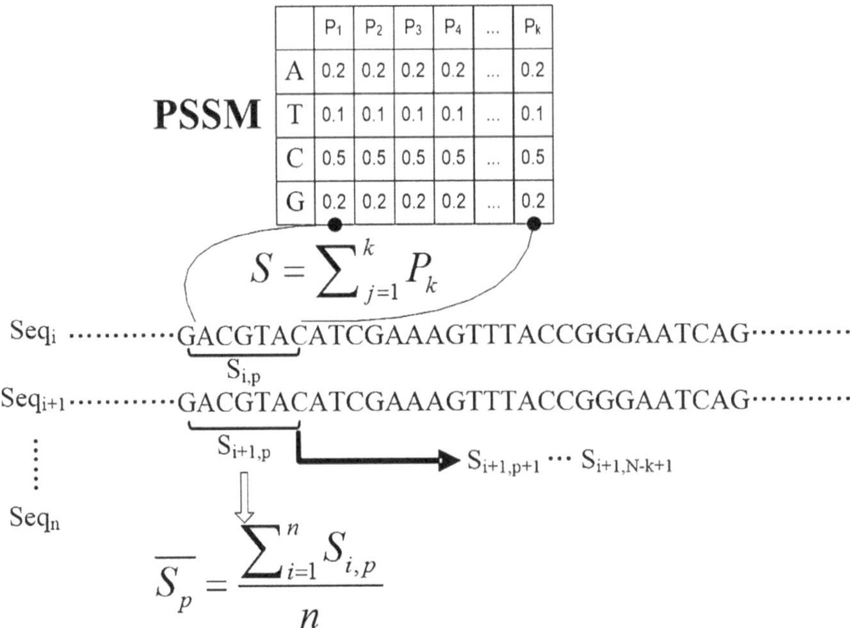

Fig. 2 Calculation of PSSM for a motif

To evaluate the statistical significance of the signals, Z-Score can be used to inspect the signals/motifs detected by Regulatory Sequence Analysis Tools (RSAT), which is based on a Markov chain model [13]. Considering the short length of triplets and tetramers, order-1 Markov Model could be used and the cutoff value for a valid Z-Score is set to 5; otherwise, order-3 Markov Model could be used and the cutoff value is 3. For the nucleotide composition in the cleavage region around poly(A) sites, Weblogo3.0 can be used to examine the profiles of nucleotide composition [14].

3.2 Single Nucleotide Profiles

To characterize the poly(A) signals, plotting the nucleotide profiles around poly(A) sites has been proven effective in identifying important sequence trends and probable cis-elements.

Given a FASTA file storing the sequences with the same length, using PAS_kpssm.pl to calculate the relative base composition of the sequences. This script mainly contains the following options:

seqfile: specify the input FASTA file.

seqdir/pat: specify the files in directory "seqdir" whose filename contains pattern "pat".

from: the start position.

to: the end position.

k: k-gram length (the length of the patterns searching for).

kfile: specify the *k*-gram file.

cnt: T/F(default); when cnt = T, then output the count of sequences.

freq: T/F(default); when freq = T, then output frequency.

tran: T/F(default); when tran = T, then transpose the output matrix.

suffix: suffix for output files.

You can use the following command to calculate the relative base composition of the sequence:

```
PAS_kpssm.pl -seqdir "seq_file_dir" -pat
"pattern" -from 1 -to 400 -k 1 -sort F -cnt
T -freq T -tran T -suffix _atcg
```

3.3 Searching Short Sequence Motifs

Before search for poly(A) signal motifs, the cis-elements need to be defined (*see* **Note 1**). SignalSleuth2 can be used to exhaustively search of such cis-elements as short sequence motifs (*k*-gram) (*see* **Note 2**) in specified range of nucleotide sequences with variable motif sizes (generally 3–8 nt in length) and rank the detected motifs based on their occurrence frequencies. Given a FASTA file storing the sequences with the same length, using SignalSleuth2.pl to calculate the relative base composition of the sequences. This script mainly contains the following options:

seqfile: to specify the input sequence file in FASTA format.

seqdir: to specify the directory containing sequence files.

pat: to specify the pattern for sequence filenames. If it is not specified or "", then the program will process all files in seqdir. Use "\.fas$" to specify all files with extension ".fas".

gap: to specify the gap value for "gap" mode. If gap > 0, then the "gap" mode is used (*see* **Note 3**).

once: T or F(default). To specify whether uses "once" mode. If once = T, then the "once" mode is used (*see* **Note 4**).

sort: T or F(default). To specify whether to sort the output. If sort = T, then sort the motif occurrence file (.cnt) by the total occurrences of each position of each motif and sort pssm file (.pssm) by max score of the given region of each motif (*see* **Note 5**).

topn: To specify the output number of top-rank patterns.

suffix: To specify the extension for the output occurrence file and pssm file. If not specified, then the output filename will be generated using the options, like <input file name>_<from>to<to>_k<k>_top<topn>_sort.cnt or .pssm, like "test1.fas_265to290_k6_top50_sort.cnt".

cnt: T or F(default). To specify the output occurrence file and pssm file.

pssm: T or F(default). To specify the output pssm file.

3.4 Examples of Running SignalSleuth2

1. Example of using the "overlapping" mode: (*see* **Note 6**)

    perl signalsleuth2.pl -seqfile ./ss2_input/test1.fas -from 1 -to 50 -k 6 -sort T -topn 50 -cnt T -pssm T -suffix "_normal"

    This command line will process a sequence file (-seqfile ./ss2_input/test1.fas); rank the hexamers (-k 6) in region 1–50 (-from 1 -to 50); output both occurrence file and pssm file (-cnt T -pssm T); sort the results by occurrence and pssm score (-sort T); only output the top 50 *k*-grams (-topn 50); use the suffix "_normal" for the filenames of output files. Finally, two files "test1.fas_normal.cnt" and "test1.fas_normal.pssm" will be generated.

2. Example of using the "gap" mode (*see* **Note 7**).
    Use "-gap 2" to specify the gap value for "gap" mode. Other options just like example 1.

    perl signalsleuth2.pl -seqfile ./ss2_input/test1.fas -from 1 -to 50 -k 6 -gap 2 -sort T -cnt T -pssm T -topn 50 -suffix "_gap"

3. Example of using the "once" mode (*see* **Note 6**).
    Use "-once T" to specify the "once" mode. Other options just like example 1 and 2.

    perl signalsleuth2.pl -seqfile ./ss2_input/test1.fas -from 1 -to 50 -k 6 -once T -sort T -topn 50 -cnt T -pssm T -suffix "_once"

4. Specify a directory containing multiple input FASTA files. This command line will process all files whose filename contains "arab" in directory "testdata" under current directory.

5. Finally, the output occurrence file and pssm file of each input fasta file named <input filename>kp.cnt and <input filename>kp.pssm will be generated.

    perl signalsleuth2.pl -seqdir testdata -pat "arab" -from 1 -to 40 -k 6 -gap 2 -sort T -cnt T -pssm T -topn 50 -suffix "kp"

4 Notes

1. Definition for poly(A) signal elements: Polyadenylation signals are cis-elements surrounding the cleavage sites [or poly(A) sites] that are recognized by the polyadenylation complex and direct both the cleavage and polyadenylation reactions. Until now there is no precise way to define the locations for poly(A) signal elements (e.g., FUE, NUE, and CE). However, according to previous researches [1, 8, 15–18], the NUE region could be defined using the following two criterions: (1) single

nucleotide profile: NUE could start around the first crossing site of A and U (around −30) and end around the another crossing site of A and U; (2) the canonical motifs (if any, like AAUAAA) should exist in this region. Once NUE is determined, FUE will be a range immediately upstream of NUE, which could be defined based on the single nucleotide profile. The start position of FUE is normally defined as the position where dominant G or U should appear. If not, an arbitrary position like the upstream 200 nt could be used as the FUE start position. Another signal region CE is the region around poly(A) site, the start and end positions of which could normally be defined as the −10 and +10.

2. Three k-gram scanning modes were used to filter out useless k-grams: overlapping mode, gap mode and once mode. Overlapping mode is the simplest mode to scan the studied region and obtain the number of occurrences of all k-grams. In the once mode, each k-gram is counted once and the last location it appearing in the studied region is recorded. The gap-mode can be adopted to prevent counting twice for mutually overlapping occurrences.

3. Occasionally, a target motif may appear multiple times within a given region of a sequence, resulting in over-representation of a specific motif. SignalSleuth2 provides the gap scanning mode to count non-overlapping signal frequency in a given region for a sequence, where a distance parameter (-gap) could be specified to prevent over-counting of these motifs. For example, GCGCGC will be counted only once in sequence … GCGCGCGC… if setting gap = 5.

4. In the cases where there is more than one occurrence of non-overlapped motif in a given region, SignalSleuth2 provides the "once" scanning mode to choose the motif that is the closest to the poly(A) site. If a pattern appears more than once in the given region of one sequence, then it is only counted once and the last position is considered as its final position. Note that, you cannot specify both "-gap" and "-once" in one command line. If "gap < 0" and "once = F", then the "overlapping" mode will be used.

5. For each scanning mode, the SignalSleuth2 provides PSSM results simultaneously. The PSSM score was calculated based on the number of the occurrences of each candidate pattern. The PSSM matrix has four rows and k columns, corresponding to the four bases (A, T, C and G) and the length of k-gram, respectively. For a given sub-sequence with the length equal to the column number of the PSSM matrix, its score is the sum of individual scores at all nucleotide positions. Higher score indicates the higher likelihood of the presence of a pattern similar to the k-gram represented by the PSSM matrix.

6. Normally, given a short region like the NUE where most k-grams only appear once, we can use the simplest overlapping-mode to scan the studied region and obtain the number of occurrences of all k-grams. Whereas if a k-gram appears more than once in a given region of one sequence, the once-mode can be adopted to decide whether or not this k-gram is present, where each k-gram is counted once and the last location it appearing in the studied region is recorded.

7. SignalSleuth2 provides the gap-mode for calculating the number of occurrences of k-grams in long sequences (e.g., 400 nt) to prevent overlapping matches for some periodic k-grams like GCGCGC. Generally, for short sequences, these three modes return similar results when scanning k-grams with a certain length ($k \geq 5$). While for long sequences, the once-mode or gap-mode may be more appropriate in that they can avoid counting too many times for mutually overlapping occurrences, especially for a stretch of the same nucleotide like TTTTTTn.

Acknowledgement

Funding supports for this work were from the National Natural Science Foundation of China (Nos. 61174161 and 61304141), the Natural Science Foundation of Fujian Province of China (No. 2012J01154), the specialized Research Fund for the Doctoral Program of Higher Education of China (Nos. 20130121130004 and 20120121120038), and the Fundamental Research Funds for the Central Universities in China (Xiamen University: No. 2013121025), Xiamen Shuangbai Talent Plan (to QQL), and US National Science Foundation (grant nos. IOS–0817829 and IOS-1353354 to QQL).

References

1. Hu J, Lutz CS, Wilusz J, Tian B (2005) Bioinformatic identification of candidate cis-regulatory elements involved in human mRNA polyadenylation. RNA 11(10):1485–1493

2. Bartel DP (2009) MicroRNAs: target recognition and regulatory functions. Cell 136(2):215–233. doi:10.1016/j.cell.2009.01.002

3. Wickens M, Bernstein DS, Kimble J, Parker R (2002) A PUF family portrait: 3′ UTR regulation as a way of life. Trends Genet 18(3):150–157

4. Holec SLH, Kuhn K, Alioua M, Borner T, Gagliardi D (2006) Relaxed transcription in Arabidopsis mitochondria is counterbalanced by RNA stability control mediated by polyadenylation and polynucleotide phosphorylase. Mol Cell Biol 26:2869–2876

5. Hammell CMGS, Zenklusen D, Heath CV, Stutz F, Moore C, Cole CN (2002) Coupling of termination, 3′ processing, and mRNA export. Mol Cell Biol 22:6441–6457

6. Buratowski S (2005) Connections between mRNA 3′ end processing and transcription termination. Curr Opin Cell Biol 17:257–261

7. Moor CH, Meijer H, Lissenden S (2005) Mechanisms of translational control by the 3′ UTR in development and differentiation. Semin Cell Dev Biol 16(1):49–58. doi:10.1016/j.semcdb.2004.11.007

8. Beaudoing E, Freier S, Wyatt JR, Claverie JM, Gautheret D (2000) Patterns of variant polyadenylation signal usage in human genes. Genome Res 10(7):1001–1010

9. Legendre M, Gautheret D (2003) Sequence determinants in human polyadenylation site selection. BMC Genomics 4(1):7

10. Liu H, Han H, Li J, Wong L (2003) An in-silico method for prediction of polyadenylation signals in human sequences. Genome Inform 14:84–93

11. Akhtar MN, Bukhari SA, Fazal Z, Qamar R, Shahmuradov I (2010) POLYAR, a new computer program for prediction of poly(A) sites in human sequences. BMC Genomics 11(1):646

12. Cheng Y, Miura RM, Tian B (2006) Prediction of mRNA polyadenylation sites by support vector machine. Bioinformatics 22(19): 2320–2325

13. Thomas-Chollier M, Sand O, Turatsinze JV, Janky R, Defrance M, Vervisch E, Brohee S, van Helden J (2008) RSAT: regulatory sequence analysis tools. Nucleic Acids Res 36(Web Server):W119–W127. doi:10.1093/nar/gkn304

14. Crooks GE, Hon G, Chandonia JM, Brenner SE (2004) WebLogo: a sequence logo generator. Genome Res 14(6):1188–1190. doi:10.1101/gr.849004

15. Ji G, Zheng J, Shen Y, Wu X, Jiang R, Lin Y, Loke JC, Davis KM, Reese GJ, Li QQ (2007) Predictive modeling of plant messenger RNA polyadenylation sites. BMC Bioinform 8(43):43

16. Loke JC, Stahlberg EA, Strenski DG, Haas BJ, Wood PC, Li QQ (2005) Compilation of mRNA polyadenylation signals in Arabidopsis revealed a new signal element and potential secondary structures. Plant Physiol 138(3): 1457–1468

17. Shen Y, Ji G, Haas BJ, Wu X, Zheng J, Reese GJ, Li QQ (2008) Genome level analysis of rice mRNA 3′-end processing signals and alternative polyadenylation. Nucleic Acids Res 36(9):3150–3161

18. Shen Y, Liu Y, Liu L, Liang C, Li QQ (2008) Unique features of nuclear mRNA poly(A) signals and alternative polyadenylation in Chlamydomonas reinhardtii. Genetics 179(1): 167–176

Chapter 2

Prediction of Plant mRNA Polyadenylation Sites

Xiaohui Wu, Guoli Ji, and Qingshun Quinn Li

Abstract

Messenger RNA polyadenylation is one of the essential processing steps during eukaryotic gene expression. The site of polyadenylation [poly(A) site] marks the end of a transcript, which is also the end of a gene in most cases. A computation program that is able to recognize poly(A) sites would not only be useful for genome annotation in finding genes ends, but also for predicting alternative poly(A) sites. PASS [*Poly(A) Site Sleuth*] and PAC [*Poly(A) site Classifier*] were developed to predict poly(A) sites in plants. PASS was built based on the Generalized Hidden Markov Model (GHMM), which consists of four functional modules: input model, poly(A) site recognition module, graphic process module, and output module. PAC is a classification model, integrating several features that define the poly(A) sites including K-gram pattern, Z-curve, position-specific scoring matrix, and first-order inhomogeneous Markov sub-model. PAC can be used to predict poly(A) sites from species whose polyadenylation profile is unknown. The result of PASS and PAC is an output of a few files with one of them containing the score or probability of being a poly(A) site for each position of a given sequence. While the models were built mostly based on poly(A) profile data from Arabidopsis, it is also functional in other higher plants since their profiles are quite similar.

Key words Classification based modeling, Polyadenylation, Predictive modeling, GHMM, PASS, PAC

1 Introduction

The location of a terminal nucleotide of the 3′-untranslated region (3′-UTR), which is exposed after an endonuclease cleavage, is known as a poly(A) site. A poly(A) site marks the end of the transcribed mature mRNA, and, as such, it can be used to find and annotate the end of a gene in most cases. The location of a poly(A) site for a gene is mostly predetermined by the so-called polyadenylation signals. Traditionally, poly(A) sites are identified by examining the expressed sequence tags (ESTs) which are reverse-transcribed from mature mRNA. Since the poly(A) tail is added post-transcriptionally, alignment of ESTs to their respective genomic sequences will reveal the location of poly(A) sites. Indeed, there are a number of datasets with collections of poly(A) sites [1–5]. Further analyses of these

Arthur G. Hunt and Qingshun Quinn Li (eds.), *Polyadenylation in Plants: Methods and Protocols*, Methods in Molecular Biology, vol. 1255, DOI 10.1007/978-1-4939-2175-1_2, © Springer Science+Business Media New York 2015

13

datasets have elucidated the poly(A) signals that determine the poly(A) site locations at the genome level. Such information about poly(A) signals, particularly those from Arabidopsis [3] and rice [5], is the foundation of this work which involves the construction of predictive models for a systematic prediction of plant poly(A) sites.

The complexity of poly(A) sites is demonstrated by the fact that poly(A) sites can be located in a short region of the 3′-UTR [3, 5, 6]. Furthermore, plant poly(A) signals possess little conservation. These properties, coupled with a limited knowledge of numerical prediction and its application to plant polyadenylation, make it hard to predict exact poly(A) sites using computational methods. PASS [7] and PAC [8] were developed to predict poly(A) sites in plants, which were based on generalized hidden Markov model (GHMM) and a classification-based prediction model, respectively. PASS consists of four functional modules: input model, poly(A) site recognition module, graphic process module, and output module. In the input module, sequences or sequences with PASS predicted scores can be entered by universal FASTA format or manually. The input system will also examine the authenticity of the sequences in that warning will be given when undesirable characters other than A, T, G, C, or U (the five nucleotides found in DNA and/or RNA) is entered. As the core of the algorithm, the function of the poly(A) site recognition module is to implement the forward-backward algorithm of the GHMM-based poly(A) site recognition model, and produces a probability score that each of the nucleotides to be a potential poly(A) site. In the graphic processing module, given the output value (or scores) from the site recognition module, PASS is designed to show the numerical scores and other results in a graphic display. Moreover, users may choose their desire color and line styles etc. The graphs can be saved as pictures in .bmp or .wmf formats. In the output module, the poly(A) site recognition results are saved in a format that allows traceability. Thus, the results as a whole (including scores, Sensitivity, and Specificity) are saved for future reference.

Another poly(A) site prediction tool, PAC, was designed to extract poly(A) sites from known dataset (with poly(A) specified) and use this information to predict poly(A) sites from unknown sequences. This is powerful when dealing with a new species that is substantially different (or do not know if there is different) from Arabidopsis or rice models. It constitutes three major steps. In the first step, PAC takes a set of nucleotide sequences as input and imports the parameter file for the classification model and the features. In the second step, a classification algorithm is performed to train and test the model. In the final step, the output files are generated. The output of PAC is a few text files with one of them containing the probabilities of each sequence to determine the potential poly(A) sites and the others as intermediate files.

While originally designed to identify the poly(A) sites in Arabidopsis, PASS or PAC may be adapted for species other than Arabidopsis if the species in question has sufficient number of poly(A) sites to train the model parameters. PASS and PAC had been successfully applied to identify the poly(A) sites in Arabidopsis [7, 8], rice [4] and green alga *Chlamydomonas reinhardtii* [9]. In this chapter, we use Arabidopsis data set as an example to illustrate how to use PASS and PAC to identify poly(A) sites.

2   Materials

2.1   Equipment

Computer Hardware: MS Windows XP/Vista/7 or later.

To use PAC, please make sure that Weka and Java VM are installed in your system. You can download a self-extracting executable version of Weka that includes Java VM 5.0 from http://prdownloads.sourceforge.net/weka/weka-3-5-6jre.exe. Other versions of Weka can be downloaded from http://www.cs.waikato.ac.nz/ml/weka/.

2.2   Data

The 8 K dataset or other data sets can be downloaded from the following links: http://www.users.miamioh.edu/liq/links.html; or http://www.polyA.org.

2.3   Setting Up

Download PASS or PAC packages from the following links: http://www.users.miamioh.edu/liq/links.html; or http://www.polyA.org.

2.4   Install PASS or PAC on MS Windows

Simply unzip the PASS or PAC package to any directory.

3   Methods

3.1   Using PASS to Predict Poly(A) sites

3.1.1   Generalized Hidden Markov Model Used in PASS

PASS was developed based on the GHMM. The topological structure was one of the most important factors in designing a GHMM model. PASS employed a GHMM model that recognized the signals from left to right, and only allowed the recognition of signals from the current stage to the next stage in one direction. Based on the analysis of the model of plant poly(A) signals [3], the sequences can be classified into five regions (Fig. 1a). The poly(A) signals are distributed in these regions with some spacing between the two signal elements. Based on this, a background stage was added between the two signal stages. To simplify the model, the length of every signal was fixed but the length of background was variable. It was also possible that two signals were next to each other and thus the length of the background may be 0. The final model was designed in such a way that all calculations began on the first stage and ended at the last stage (Fig. 1b).

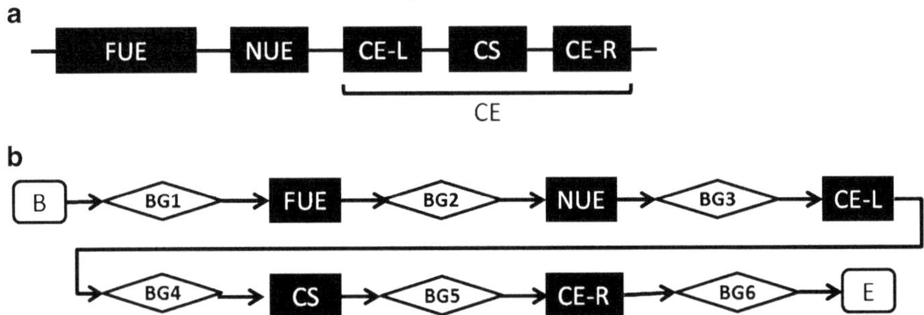

Fig. 1 The structure of plant mRNA polyadenylation signals, the order of the GHMM, and a flowchart of PASS. (**a**) A working model. (**b**) The order of GHMM. The *arrowheads* indicate the probability of changing of states (all probabilities were set to be 1). The *rectangles* represent regions with fixed length while the *braces* indicate regions with variable length. *FUE* far upstream element, *NUE* near upstream element, *CE* cleavage element, *CE-L*, *CE-R* cleavage element left or right to the poly(A) site, *CS* cleavage site, also known as poly(A) site, *B* beginning of the scan, *Bg* background sequences between cis-elements, *E* end of the scan. Note that because YA is not found in all sequences, other dinucleotide combinations are also considered in GHMM

In this model, some basic parameters were set as follows: the number of stages was 11 (Fig. 1b, from Bg1 through Bg6); the array of signals in every stage was set to be {A, T, C, G}. Stages in odd numbers were the background stage with variable length, and stages in even numbers were signal stages with fixed length. Because the model begins with the first stage and ends at the last stage, the initial state distribution was set in $\pi = \{1, 0, ..., 0\}$. Every stage (i) can be only transferred to the $i+1$ stage. Each stage was assigned a few parameters including the signal nucleotide composition, signal pattern length, etc.

A sliding 180 nt-wide window was applied to calculate the outputs of scores for the sequences. For every nucleotide, PASS deduced a score in all windows that contained this nucleotide. The window slides along the entire sequence, combining values of forward-backward variables to generate a score value for each nucleotide in that window.

3.1.2 Interface of PASS

The interface of PASS mainly consists of a title bar, a menu bar, a graph display area, a sequence display area, and an analysis operation area, as shown in Fig. 2. Furthermore, PASS possesses the function of language switching (between English and Chinese), which is very convenient for people using these two languages. All the terms and concepts in captions of PASS rest on the knowledge about bioinformatics and the experience from biology researchers.

3.1.3 Import One or More Sequences from File

Choose the menu "File → Import sequence only". The format of the sequences is FASTA, and PASS omits blank lines. If you already have the output file of PASS and want to display the result graphically, you may choose "Import Sequences with Score From File", which is almost the same as import sequences without scores from file.

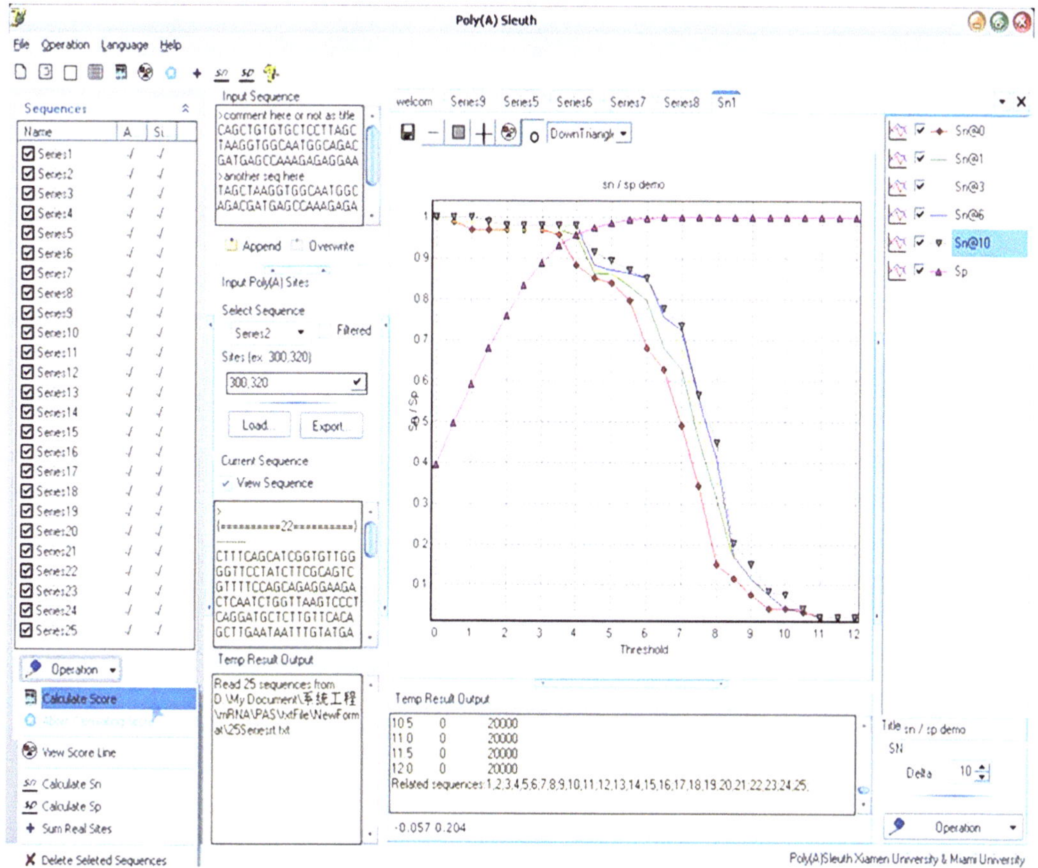

Fig. 2 Screenshot of the interface of PASS

3.1.4 *Calculate Scores*	Choose the menu "Operation → Calculate Score". During the calculation, the marquee status bar will show "calculating seriesX", which means seriesX is being calculated. When a sequence has been calculated, the corresponding place of column [A..] will be marked as "$\sqrt{}$". To display the sequence graphically, choose "Operation → View Score Line" or double click a sequence.
3.1.5 *Evaluation of the Results from the Model*	Before evaluation, you need to give known poly(A) sites in the sequences. To input the known poly(A) sites, you need to switch to the [Input Poly(A) Sites] Panel. You may load the poly(A) sites from a file with the numbers indicating the positions of the sites or manually input the known sites. Then you can choose "Operation → Evaluate(Sn,Sp)" to calculate the sensitivity (Sn) and specificity (Sp) (*see* **Note 1**).
3.1.6 *Related Operations*	1. *Calculate Score*: calculate score of the selected sequences.
	2. *Abort Calculate Score*: abort Calculating Score of the selected sequences. This action will abort the calculation of all the selected sequences.

3. *View Score Line*: view the score line of the selected sequence. Double click the sequence can have the same effect.

4. *Calculate Sn*: calculate Sn and Sp using the selected sequences.

5. *Calculate Sp*: calculate Sp separately. You can specify your own random sequences.

6. *Calculate Sp of Selected Sequences*: calculate Sp of selected Sequences. You must select some calculated sequences first.

7. *Sum Real Sites*: calculate the total count of the sites of the selected sequences. The result will be shown in [Temp result Output] Panel (*see* **Note 2**).

3.1.7 An Example to Predict Poly(A) Sites

1. Input Sequences
   method1:[File] → [Import] → [Import sequences only].
   method2:manually input sequences from [Input Sequences] panel.

2. Calculate Score
   Choose "[Operation] → [Calculate Score]" (*see* **Note 3**).

3. View Score Line
   Choose "[Operation] → [View Score Line]" or double click a sequence.

4. Input Poly(A) Sites
   Use "[Input Poly(A) Sites]" Panel. You can load sites from a file or manually input the sites.

5. Calculate Sn and Sp
   Choose "[Operation] → [Evaluate(Sn,Sp)]" to calculate Sn and Sp.

3.2 Using PAC to Predict Poly(A) Sites

3.2.1 Classification-Based Model Used in PAC

When using a classification algorithm for predicting poly(A) sites, the nucleotide sequence needs to be converted into numeric format. Consequently, the features of poly(A) signals around the cleavage sites were extracted based on the profile of nucleotide sequence distribution around poly(A) sites. In PAC, five feature representation methods were adopted to describe the makeup of nucleotide sequences. These methods were chosen to confirm whether each one could generate unique features from different training datasets. Finally, the numerical vector was used as the input of the classification algorithm. The distribution of features in different areas of a window sequence is shown in Fig. 3. There are several kinds of feature representation methods in PAC, including K-gram nucleotide sequence pattern, Z-Curve, the position-specific scoring matrix based score, the probability based on first order inhomogeneous Markov sub-model, and the weight of NUE Signals.

To find the feature group that is most effective for poly(A) site prediction, the selection algorithm involves feature evaluation and searching algorithms. Six commonly used feature evaluation methods

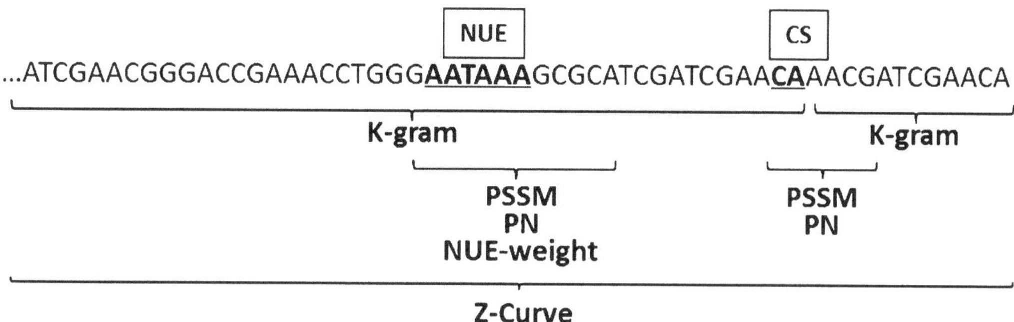

Fig. 3 Different methods were used to extract features from different regions of the poly(A) signals. *NUE* near upstream element, *CS* cleavage site, *PSSM* position-specific scoring matrix, *PN* probability based on first-order inhomogeneous Markov sub-model

were adopted to select the relatively optimal subset of features. These methods include ChiSquaredAttributeEval, GainRatioAttributeEval, InfoGainAttributeEval, OneRAttributeEval, ReliefFAttributeEval, and SymmetricalUncertAttributeEval, and four searching algorithms, including ExhaustiveSearch, GeneticSearch, GreedyStepwise, and RandomSearch, in an open source software called WEKA3.5 [10].

The classification-based poly(A) site recognition model (PAC) is based on two steps: training and testing. For the training classification model, the window sequences in the training dataset were converted into numeric form using the final feature-space. Then, the numeric formatted sequences were classified using one of the classification algorithms to build a training model. Different classification algorithms were used to build different training models. In the testing step, the training model was used for predicting the poly(A) sites from the test dataset. As each given sequence passed through the test model, each position of the sequence was predicted as a poly(A) site, or not.

Long sequence analysis involved a pre-processing step in which a given length, $n - 162 + 1$ (where n is the length of the sequence) of sub-sequences, was first produced according to the window size before entering into the model. Then, the final output of the model could be shown as the prediction results of the sequences. Both the training and test processes permit inputting sequences longer than 162 nt; however, the long sequences were manually cut into 162 nt window sequences as inputs for the training process.

3.2.2 Interface of PAC

The interface of PAC mainly consists of a title bar, a menu bar, several panels for sequence input, Weka settings, Feature settings, and sequence output.

3.2.3 Introduction About the Related Files

1. Ftr File(.ftr) The file integrating all settings of wanted features.
2. Arff File(.arff) The file for WEKA transformed from the data matrix.
3. Model File(.model) The output file of training model needed in the test model.
4. Matrix File(.mtx) The file of a data matrix.
5. Pr File(.pr) The output file of WEKA.
6. Pred File(.pred) The file transformed from the pr File.

3.2.4 Sequence Input

Two ways are allowed for sequence input. One way is to input the "True/False sample" files, which inputs the short sequences with or without poly(A) sites, respectively. Another way is to input the "long sequences/sites" files, which specifies the long sequences with multiple sites and the sites file.

3.2.5 Weka Settings

Click the "weka.jar" button to specify the file path of weka.jar. If weka is installed in your windows, PAC will get the file path automatically. If the checkbox "-p distribution" is checked, the pr file will contain the probability of the classification, otherwise the confusion matrix.

3.2.6 Output Settings

When the checkbox "Continuous Output" is checked, the Ftrs/Arff/Model files will be generated if the Ftrs/Arff/Model files are specified. When it is unchecked, only the final file is generated.

When the checkbox "Optimal Output" is checked, PAC will not calculate the matrix but use the arff file specified to generate other output files (e.g., arff, model file). If it is unchecked, PAC will not use the specified files but calculate all related files in the whole process.

3.2.7 Features Settings

Set the parameters for the five features integrated in PAC, including K-gram, Markov sub-model, PSSM, Z-curve, and Weight of NUE.

3.2.8 An Example to Predict Poly(A) Sites

1. Model training (Fig. 4): Input the sequences for training and import the .ftr file to specify the feature parameters. Then click "output [.model]" button to output model files.
2. Model testing (Fig. 5): Input the sequences for testing and import the .ftr file to specify the feature parameters. Specify the model files from **step 1**. Then click "output [.pred]" button to output the prediction files.

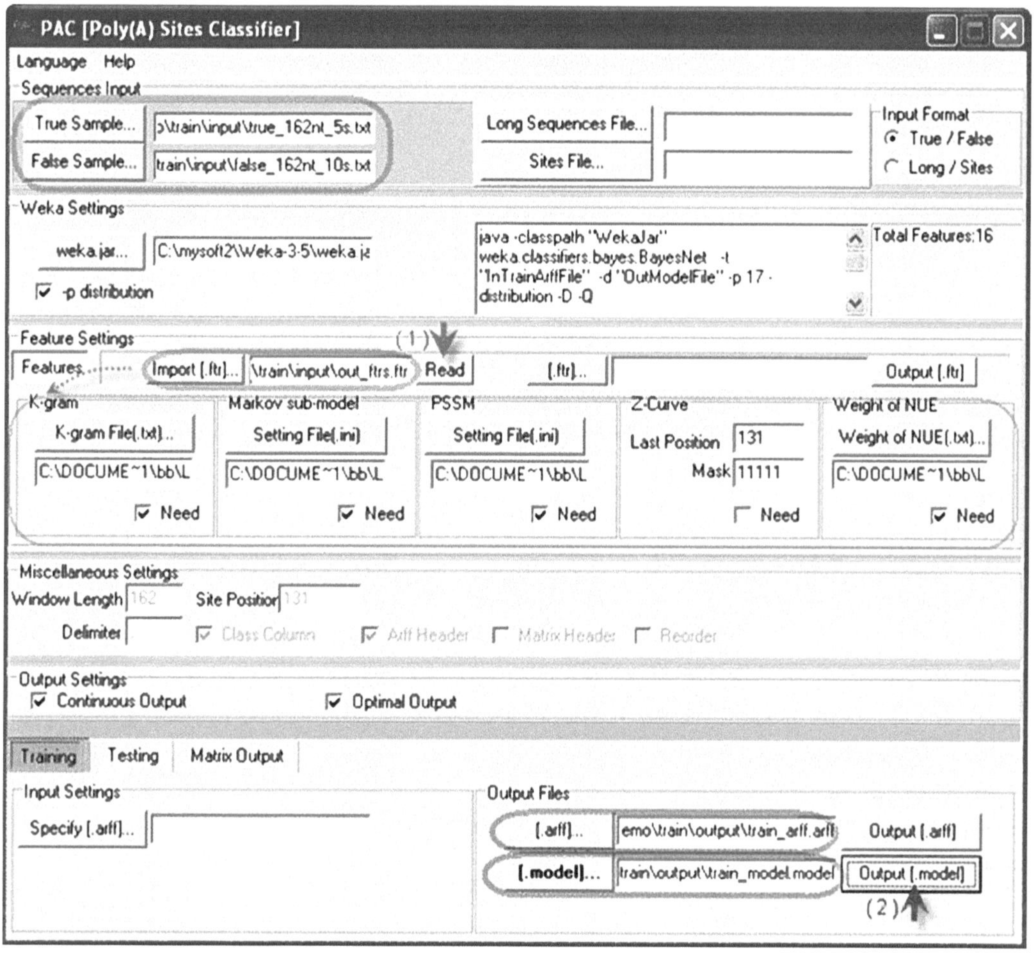

Fig. 4 Screenshot of the training step of PAC

4   Notes

1. You can calculate Sp without any specified sequence file or poly(A) sites. This operation requires a relatively long sequence and it is already calculated (i.e., the specified file must have both the sequence and scores). If there is any sequence in selection whose background color is in grey, it will be used in calculation of Sp and a new line will be created. Otherwise, a blank page is created, then you can specify a file to calculate Sp. Choose "Operation → Calculate Sp" to calculate Sp.

2. Except the action of view score line and calculate Sp, other operations are only in effect when some sequences are selected. That is, before these actions, you need to selected sequences from [sequences] panel.

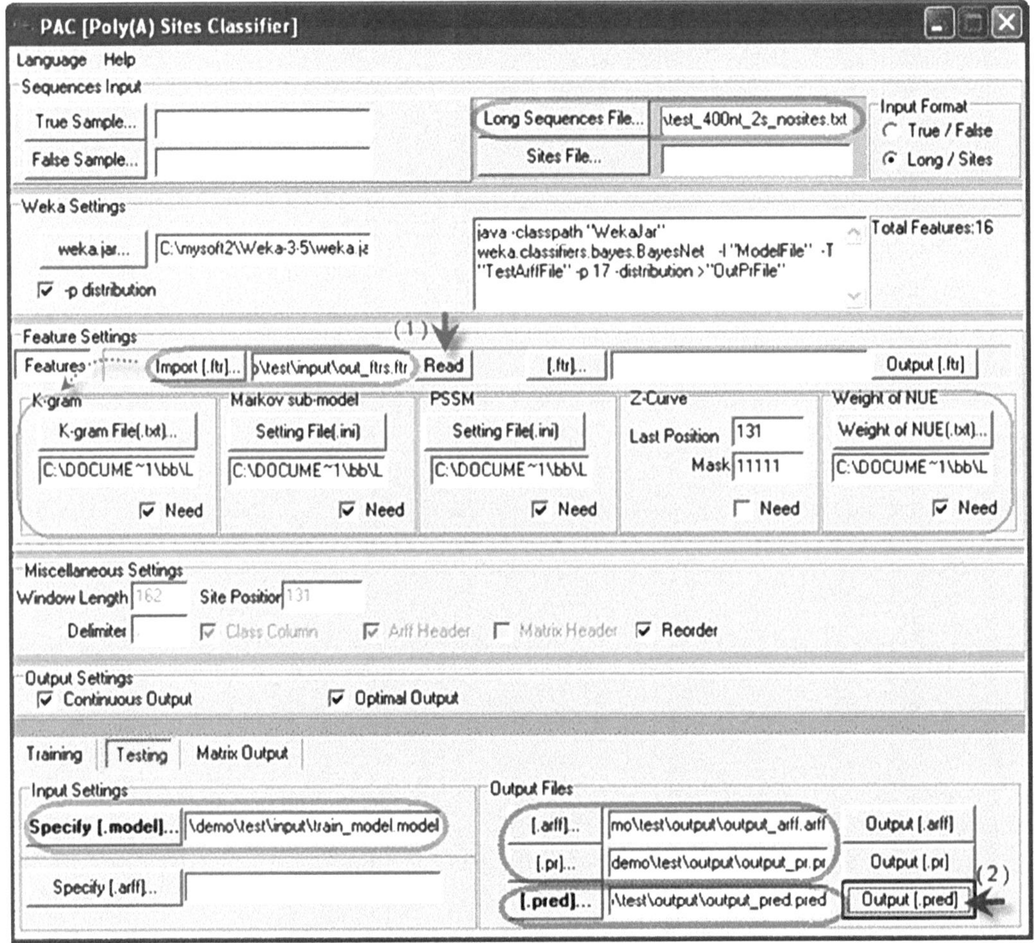

Fig. 5 Screenshot of the testing step of PAC

3. In the calculation, the marquee status bar will show "calculating seriesX", which means seriesX is being calculated. When a sequence has been calculated, the corresponding place of column [A..] will be marked as "$\sqrt{}$".

Acknowledgement

Funding supports for this work were from the National Natural Science Foundation of China (Nos. 61174161 and 61304141), the Natural Science Foundation of Fujian Province of China (No. 2012J01154), the specialized Research Fund for the Doctoral Program of Higher Education of China (Nos. 20130121130004 and 20120121120038), and the Fundamental Research Funds for the Central Universities in China (Xiamen University: No. 2013121025), Xiamen Shuangbai Talent Plan (to QQL), and US National Science Foundation (grant nos. IOS–0817829 and IOS-1353354 to QQL).

References

1. Graber JH, Cantor CR, Mohr SC, Smith TF (1999) In silico detection of control signals: mRNA 3′ end-processing sequences in diverse species. Proc Natl Acad Sci U S A 96(24): 14055–14060

2. Graber JH, Cantor CR, Mohr SC, Smith TF (1999) Genomic detection of new yeast pre-mRNA 3′-end-processing signals. Nucleic Acids Res 27(3):888–894

3. Loke JC, Stahlberg EA, Strenski DG, Haas BJ, Wood PC, Li QQ (2005) Compilation of mRNA polyadenylation signals in Arabidopsis revealed a new signal element and potential secondary structures. Plant Physiol 138(3):1457–1468

4. Shen Y, Ji G, Haas BJ, Wu X, Zheng J, Reese GJ, Li QQ (2008) Genome level analysis of rice mRNA 3′-end processing signals and alternative polyadenylation. Nucleic Acids Res 36(9):3150–3161

5. Shen Y, Liu Y, Liu L, Liang C, Li QQ (2008) Unique features of nuclear mRNA poly(A) signals and alternative polvadenylation in Chlamydomonas reinhardtii. Genetics 179(1): 167–176

6. Li QQ, Hunt AG (1997) The polyadenylation of RNA in plants. Plant Physiol 115:321–325

7. Ji G, Zheng J, Shen Y, Wu X, Jiang R, Lin Y, Loke JC, Davis KM, Reese GJ, Li QQ (2007) Predictive modeling of plant messenger RNA polyadenylation sites. BMC Bioinform 8(43):43

8. Ji G, Wu X, Shen Y, Huang J, Li QQ (2010) A classification-based prediction model of messenger RNA polyadenylation sites. J Theor Biol 265(3):287–296. doi:10.1016/j.jtbi. 2010.05.015

9. Ji G, Wu X, Li Q, Zheng J (2010) Messenger RNA polyadenylation site recognition in green alga Chlamydomonas Reinhardtii. Lect Notes Comput Sci 6063:17–26. doi:10.1007/ 978-3-642-13278-0_3

10. Witten IH, Frank E (2005) Data mining: practical machine learning tools and techniques. Elsevier, San Francisco, CA

Chapter 3

Extraction of Poly(A) Sites from Large-Scale RNA-seq Data

Min Dong, Guoli Ji, Qingshun Quinn Li, and Chun Liang

Abstract

The NCBI manages the SRA (Sequence Read Archive) database to store RNA-Seq data generated from different NGS technologies. With ever increasing finished and ongoing genome and transcriptome sequencing projects, the data in SRA expand rapidly and present a treasure for mining useful information to facilitate our understanding of biological issues like mRNA 3′-end formation and alternative polyadenylation. We developed a bioinformatics pipeline that can process raw SRA sequence data and obtain high quality poly(A) sites and poly(A) cluster sites with detailed expression information. This pipeline is designed to be generic and can be utilized for polyadenylation studies in any eukaryotic species.

Key words Polyadenylation, RNA-Seq, SRA, Poly(A) site, Data mining

1 Introduction

The Sequence Read Archive (SRA) at the US National Center for Biotechnology Information (http://www.ncbi.nlm.nih.gov/sra) stores raw sequence reads from next-generation sequencing technologies including 454, Illumina, Helicos, among others [1]. With the rapid advance in sequencing technologies, sequences in SRA increase dramatically in an unprecedented speed. As of October 19, 2013, there are 1.87 E+15 nucleotide bases in SRA, which is an increase of over 1,000 fold over the last 5 years (http://www. ncbi.nlm.nih.gov/Traces/sra/sra_stat.cgi). SRA contains not only sequence reads and associated quality scores, but also experimental meta-data and sometimes secondary analysis data such as alignments to the reference genomes. It covers both genomics and transcriptomics sequencing projects. Meanwhile, functional genomics studies for gene expression and regulation using RNA-Seq need to submit relevant data to Gene Expression Omnibus (GEO) (http:// www.ncbi.nlm.nih.gov/geo/). However, the original raw data files that contain sequence reads and quality scores from these studies are uploaded into SRA. In another word, SRA stores the raw sequencing data, whereas GEO archives experimental meta

Arthur G. Hunt and Qingshun Quinn Li (eds.), *Polyadenylation in Plants: Methods and Protocols*, Methods in Molecular Biology, vol. 1255, DOI 10.1007/978-1-4939-2175-1_3, © Springer Science+Business Media New York 2015

information (ecotype background, genotype/variation, tissue, development stage, organism, growth protocol, and extraction protocol), gene expression patterns, and library strategy for the studies. Users can easily download and query all relevant data by using SRA reference ID and GEO reference ID. Without a doubt, SRA is a treasure for mining raw sequence reads that contain post-transcriptional poly(A) tails because many RNA-Seq projects have sampled polyadenylated mRNAs as a way to assess the transcriptomes. Here, we present a generic bioinformatics protocol/pipeline that processes RNA-Seq raw data downloaded from SRA to determine poly(A) sites in the reference of genomic coordinates, providing a solid foundation for downstream poly(A) data analysis. While the examples are given to process Arabidopsis data, it is applicable to any species (Fig. 1).

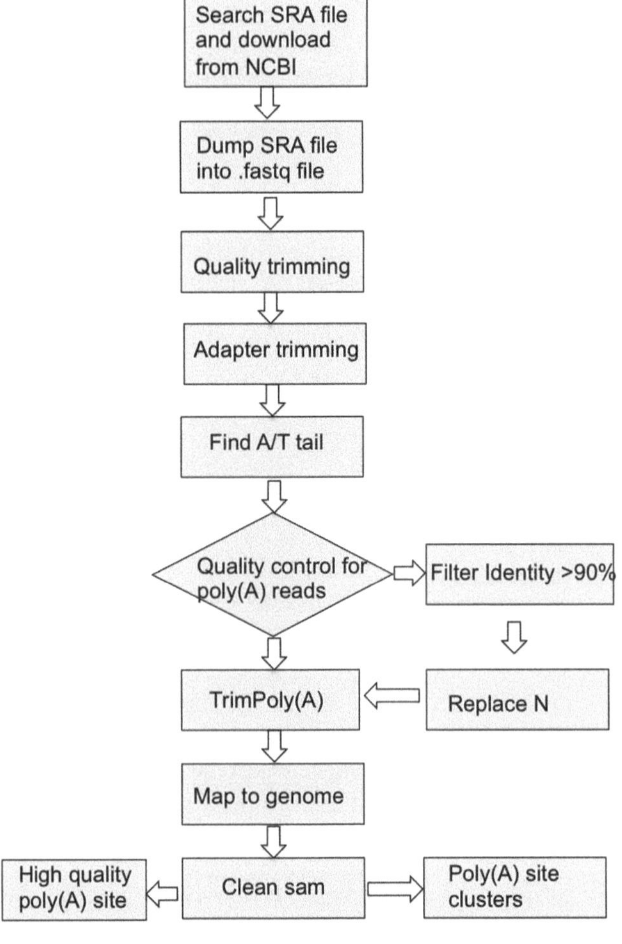

Fig. 1 The workflow used in this protocol. See text for more details

2   Materials

The software tools used for our pipeline require a 64-bit CPU computer running on open-source Ubuntu Linux OS (http://www.ubuntu.com/). This pipeline is based on pure C++ project, so it does not depend on other library. The standard Linux system environment is sufficient. The minimal amount of running RAM is depended on how big data need to be processed but we routinely use 32 GB. During processing of RNA-Seq data, at least 100 GB hard-drive space is needed. Some SRA files are very big, and extracted .fastq files may be over 50 GB. Dependent on the size of SRA files that you downloaded, more disc space might be required for a given research project.

2.1 Download and Format SRA Data Files

Originally, SRA file is based on a binary format, but it can be easily translated into .fastq file by *SRA Toolkit. Fastq-dump* is a tool of NCBI *SRA Toolkit* that can dump and extract SRA data in *.sra format from NCBI SRA into *.fastq format. *SRA Toolkit* is a freely available program, which can be downloaded from (http://eutils.ncbi.nih.gov/Traces/sra/?view=software) and installed in your Linux computer.

2.2 Quality Trimming of Raw Sequence Reads Using FASTX

High-throughput sequencing data often contain various types of errors. The low quality bases are a major problem. The sequence reads that include low quality bases must be removed. In our pipeline, quality control is performed using *FASTQ Quality Filter*, which is a tool of *FASTX-Toolkit* (http://hannonlab.cshl.edu/fastx_toolkit/download.html). *FASTQ Quality Filter* can be used to filter out low-quality sequence reads.

2.3 Adapter Trimming of Raw Sequence Reads Using FASTX

Alien sequences such as primers and adapters usually occur in both 5′end and 3′end of high-throughput sequencing data. The adapter must be found and removed from the reads before mapping. In our pipeline, adapter trimming is performed with *FASTQ/A Clipper*, which is also a program of *FASTX-Toolkit.*

2.4 Find the Poly(A) Tail

For finding the poly(A) tail in the sequencing data, we need to find the A/T rich section in each sequence read. However, the A/T rich sections may be mixed by other nucleotides. This situation makes poly(A)/(T) detection more complex. The flexible algorithm must recognize the A/T rich section and accept some minor errors within. Poly(A) tail finding is performed with our in-house program *FindTail*, which can be downloaded from (http://code.google.com/p/findtail/). The *FindTail* is a C++ program for accurate detection of poly(A)/(T) tails and other homopolymer in RNA-Seq and cDNA/EST data. It detects all perfect and imperfect poly(A) tracts in a sequence (*see* **Note 1**). User can control the error rate by input parameters (*see* in Subheading 3.2).

<table>
<tr><td>

2.5 Quality Control for Poly(A) Reads

</td><td>

The quality control for poly(A) tail is very important. Through this step, user can further control the quality of A/T tail. After *FindTail*, we are able to obtain sequence reads that contain poly(A) tails that meet our primary critical (e.g., 85 % adenine nucleotide for poly(A) tails) using *FilterIdentity*, filtering out most of other sequences that are not useful for poly(A) study (*see* **Note 2**). Next, we use *ReplaceN* to substitute ambiguous individual nucleotide N by A, T, C and G using a random algorithm in the high-quality poly(A) reads (*see* **Note 3**). Finally, *TrimPolyA* is used to trim the poly(A) tails and obtain the final clean sequence reads ready for next mRNA-to-genome mapping stage. *FilterIdentity*, *ReplaceN* and *TrimPolyA* are included in our open-source tool kit (http://code.google.com/p/polya-tools/).

</td></tr>
<tr><td>

2.6 mRNA-to-Genome Mapping

</td><td>

In this step, we require each individual clean sequence read mapped uniquely and unambiguously to the reference genome. This means that the reads mapped to multiple loci will be discarded. In our pipeline, mRNA-to-genome mapping is performed with *Helisphere* software package downloaded from the website ([2], http://sourceforge.net/projects/openhelisphere/). In addition, *Helisphere* needs SAMtools ([3]; http://samtools.sourcewforge.net/) to be installed and can generate alignment results in SAM files.

</td></tr>
<tr><td>

2.7 Quality Control of Poly(A) Sites

</td><td>

After mRNA-to-genome mapping, we are able to get all individual raw poly(A) sites supported by individual sequence reads. *CleanSam* is a Perl program that can get the true poly(A) sites in genomic coordinates in terms of the alignment results of associated mismatches, insertions and deletions. Also, it can filter out the poly(A) sites that are potential candidates for internal priming [4]. Next, *FilterPolyASite* is used to extract high quality poly(A) sites, which are defined as supported by a minimum of three individual sequence reads (*see* **Note 4**). Based on the high quality poly(A) sites, you can use *GetPolyaSiteCluster* to get poly(A) site clusters that take account of the heterogeneity issue. We have utilized advanced Ward algorithm for clusters in the *GetPolyaSiteCluster* program. *CleanSam.pl*, *FilterPolyASite* and *GetPolyaSiteCluster* are included in our open-source tool kit (http://code.google.com/p/polya-tools/).

</td></tr>
</table>

3 Methods

All programs are operated through the Linux shell scripts. Through shell script, users can process many different SRA files at the same time. This pipeline was designed with the following versions of the aforementioned programs:

- SRA Toolkit version 2.3.4-4 release.
- FASTX-Toolkit 0.0.13.
- FindTail version 1.01.

- Polya-toolkit version 1.01.
- Helisphere 0.11.
- GMAP 2012.12.12.
- SAM Tools 0.1.18.

3.1 Format of Source File and Quality Trimming of Raw Sequence Reads

1. Download the raw sequence files from NCBI SRA:

   ```
   $ wget link_name.sra
   ```

   Note: you can search your SRA id in http://www.ncbi.nlm.nih.gov/sra/, and get the download link.

2. Run fastq-dump on the *.sra files:

   ```
   $ fastq-dump name.sra
   ```

3. Quality trimming of fastq file:

   ```
   $ fastq_quality_filter -q 20 -p 90 -i input_name.fastq -o output_name.fastq
   ```

4. Trim adapter of fastq file:

   ```
   $ fastx_clipper -a adapter_string -c -i input_name.fastq -o output_name.fastq
   ```

5. Repeat **steps 1–4** for all the raw sequence files.

3.2 Find the Poly(A) Tail in Sequencing Data

1. Find the poly(A) tail in the raw sequence file:

   ```
   $ FindTail --input_file name.fastq
   --seqlength 1000 --endgap 1
   --taillength 25 --identity 85 --ptype A
   --stype T --output_format fasta
   --output_type pr --d > name.pr.pA
   ```

2. Repeat **step 1** to find the all 2 patterns tail (A/T rich section) in sequencing data (*see* **Note 5**).

Parameters for FindTail

--**input_file**: input file name, FindTail accepts *.fasta and *.fastq file.

--**seqlength**: the maximum length of your sequence in a given sequencing file.

--**endgap**: the distance between potential homopolymer and head/end of a sequence.

Note: "--endgap" parameter can control the position of A/T rich section. If user need the A/T rich section just appear at both ends, the value of "--endgap" may be set as very small (e.g., 1).

--**taillength**: the minimum length of a valid homopolymer.

--**identity**: error bases (mismatches, insertions and deletions) can be allowed in homopolymers. We use the identity parameter to control the ratio of error bases. For instance,--dentify 85 means 15 % mismatch in your homopolymer.

--**ptype**: the primary nucleotide composition in homopolymers, such as "A" is the primary composition in poly(A) tail. FindTail accepts "A" and "T" for this option.

--**stype**: when "A" is the primary composition in the first type of homopolymer and "T" should be chose for the secondary homopolymer. Both "--ptype" and "--stype" enable you to detect two different types of homopolymer in the same time (e.g., poly(A) tail and poly(T) tail).

--**output_format**: user can choose the output format between *. fasta and *.fastq.

--**output_type**: user can choose the options between pl, pr, sl, and sr (*see* **Note 6**).

--**d**: if user selects this option, the detailed information about the detected homopolymers (e.g., type=A, start=8, end=31, length=24, identity=95.8333 %) will be appended to the description line of each sequence read.

3.3 Quality Control of Poly(A) Site

1. Control the minimum identity of poly(A) tail and minimum length of the clean sequence fragment after trimming poly(A) tail:

   ```
   $ filteridentity --input_file name.pr.pA
   --seqlength 500 --identity 94
   --taillength 25 > name.pr.pA.i94
   ```

2. Replace the N character in the output file:

   ```
   $ ReplaceN --input_file name.pr.pA.i94
   --seqlength 500 --identity 94
   --Ncount 5 > name.pr.pA.RN.i94
   ```

3. Trim out the tail in the output file:

   ```
   $ TrimPolyA --input_file name.pr.pA.RN.i94
   --seqlength 500 --identity 94
   --trimside r > name.pr.pA.RN.fasta
   ```

4. To save hard disc space, remove the original .sra file:

   ```
   $ rm name.sra
   ```

5. To save hard disc space, remove the original .fastq file:

   ```
   $ rm name.fastq
   ```

6. Repeat **steps 1–5** for all the sequence files.

Parameters

--**input_file**: input file name.

--**seqlength**: the maximum length of raw sequence reads.

--**identity**: the minimum required identity for output file.

Note: Although in FindTail, "--identity" parameter already control the error rate in A/T rich section, we suggest user selecting relative lower identity value in FindTail (e.g., 85 %). User can further control the identity in this step (e.g., 90 or 95 %).

--**taillength**: the minimum required length of detected poly(A) tails.

--**Ncount**: the maximum allowed number of "*N*" characters in a given sequence read.

--**trimside**: user can choose to trim detected tails in either 5′-end or 3′-end of a given sequence. --**trimside l** means 5′-end homopolymer trimming and --**trimside r** means 3′-end homopolymer trimming.

3.4 Map Clean Reads to the Reference Genome Using Helisphere Toolkit (See Note 7)

1. Build the index files for the reference genome:

```
$ proprocessDB --reference_file
genomename.fasta --out_prefix name.seed10
```

2. Map the reads of the sample to the reference genome:

```
$ indexDPgenomic --read_file name.pr.pA.RN.fasta
--reference_file genomename.fasta
--output_file name.pr.pA.RN.bin
--data_base name.seed10
```

3. Filter alignment and control the minimum length of mapping sequence:

```
$ filterAlign --input_file name.pr.pA.RN.bin
--output_file name.pr.pA.RN.best.43.25.bin
--best_only --min_score 4.3 --min_len 25
```

Note: "--best_only" and "--min_score 4.3" control the sequence read uniquely and unambiguously mapping to the reference genome. "--min_len" controls the minimum length of sequence read in mapping.

4. Sort the alignment in the result file:

```
$ sortAlign  --input_file  name.pr.pA.RN.
best.43.25.bin
  --output_file   name.pr.pA.RN.best.43.25.
sorted.bin
```

5. Generate the .sam file:

```
$ align2sam --i name.pr.pA.RN.best.43.25.
sorted.bin
  --o name.pr.pA.RN.best.43.25.bam
```

Note: finally, user will get the mapped reads in .SAM file.

3.5 Clean SAM File Using CleanSam Program

1. clean the internal priming in sam file:

For poly(A) at 3′-end of the clean reads:

```
$perl  cleansam_pr_pa.pl  name.pr.pA.RN.
best.43.25.sam
```

For poly(T) at 5′-end of the clean reads:

```
$perl  cleansam_sl_pt.pl  name.sl.pT.RN.
best.43.25.sam
```

2. Repeat **step 1** for all SAM files (*see* **Note 8**).

<table>
<tr><td valign="top">

**3.6 Separate
the Raw Poly(A) Sites
into Different
Chromosome
and Strand (+/−)**

</td><td valign="top">

1. Combine files in the same situations into one big file.

```
$cat *.pr.pA.RN.best.43.25.sam.clean>all_
sra.pr.pA.RN.best.43.25.sam.clean
```

2. Separate the raw site into different chromosome and strand.

```
$ filtergenome all_sra.pr.pA.RN.best.43.25.
sam.clean
  --seqlength 500 --strand 0
  --chromosome_name Chr1 >
  all_sra.sam.clean_Chr1_positive.fasta
  $ filtergenome all_sra.pr.pA.RN.best.43.25.
sam.clean
  --seqlength 500 --strand 16
  --chromosome_name Chr1 >
  all_sra.sam.clean_Chr1_negative.fasta
```

3. Repeat **step 2** to all other chromosome.

</td></tr>
</table>

For the next step (cluster), we separate all raw sites into different chromosome. We use the program filtergenome, which is included in polya-toolkit.

Parameters

--**input_file**: input file name.

--**seqlength**: the maximum length of your sequence in sequencing file.

--**strand**: 0 or 16, 0 presents the positive strand, 16 presents the negative strand.

--**chromosome_name**: Chr1, Chr2, Chr3, Chr4, Chr5, ChrC, and ChrM for Arabidopsis.

Note: "--chromosome_name" depends on the chromosome name in gene annotation file, maybe some other species use the full name,such as "Chromosome1". User can verify this information in original .SAM file.

<table>
<tr><td valign="top">

**3.7 Get High-Quality
Sites Using all Raw
Poly(A) Sites**

</td><td valign="top">

1. Get high quality site in chromosome 1.

```
$  FilterPolyASite --input_file all_sra.sam.
clean_Chr1_+.fasta --seqlength 500
  --read_limit   3   >   all_sra.sam.clean_
Chr1_+_3reads.fasta
  $  FilterPolyASite --input_file all_sra.sam.
clean_Chr1_-.fasta --seqlength 500
  --read_limit 3 > all_sra.sam.clean_Chr1_-
_3reads.fasta
```

2. Repeat **step 1** to all other chromosomes.

</td></tr>
</table>

Parameters

--**input_file**: input file name.

--**seqlength**: the maximum length of your input sequences.

Table 1
An example of the poly(A) sites listed with each chromosome

A	B	C	D
Chr1	0	1188	4
Chr1	0	3565	4
Chr1	0	3567	5
Chr1	16	3973	4
Chr1	16	5672	4
Chr1	16	5675	11
Chr1	16	5703	149
...	...	...	...

--**read_limit**: the minimum number of mapped reads required to support a high-quality poly(A) site (*see* **Note 9**).

The output of this program is shown as followings (and in Table 1).

A: Chromosome name.

B: strand, 0 = positive, 16 = negative.

C: genomics coordinate of a high-quality poly(A) site.

D: supporting number of high-quality poly(A) reads for each poly(A) site.

3.8 Obtain Poly(A) Site Clusters Using Raw Poly(A) Sites for Alternative Polyadenylation Studies

1. Get poly(A) site clusters in individual chromosomes.

```
$ GetPolyaSiteCluster --input_file all_
sra.sam.clean_Chr1_+.fasta
   --seqlength 500 --read_limit 3
   --cluster_seed 25 > all_sra.sam.clean_
Chr1_positive_cluster.fasta
$ GetPolyaSiteCluster --input_file all_
sra.sam.clean_Chr1_-.fasta
   --seqlength 500 --read_limit 3
   --cluster_seed 25 > all_sra.sam.clean_
Chr1_negative_cluster.fasta
```

2. Repeat **step 1** to all other chromosome.

Parameters

--**input_file**: input file name.

--**seqlength**: the maximum length of your input sequences.

--**read_limit**: the minimum number of mapped reads required to support a high-quality poly(A) site (*see* **Note 9**).

Table 2
An example output

A	B	C	D	E	F	G	H
Chr1	5654	5680	+	pA:5680	height:502	reads:506	GAP_size:4924
Chr1	10604	10604	+	pA:10604	height:4	reads:4	GAP_size:2394
Chr1	12998	12998	+	pA:12998	height:192	reads:192	GAP_size:80
Chr1	13078	13078	–	pA:13078	height:10	reads:10	GAP_size:5178
Chr1	18256	18259	–	pA:18256	height:9	reads:18	GAP_size:2538
Chr1	20797	20797	–	pA:20797	height:4	reads:4	GAP_size:2656
…	…	…	…	…	…	…	…

--**cluster_seed**: the initiate seed value for the clustering algorithm. We recommend it to be 25 based on our experimentation. The value closer to the real value, the less the time utilized by the clustering algorithm.

The output of this program is shown as followings (and Table 2).

A: Chromosome name.

B: the start position in genomic coordinate of one poly(A) cluster.

C: the end position in genomic coordinate of one poly(A) cluster.

D: strand (+/−) of the original genome annotation.

E: the representative site of one given poly(A) cluster that spans the genomic region from B to C. This position has the highest sequence read support.

F: the support number of the representative site.

G: the support number of all high-quality poly(A) reads that support any component poly(A) sites within a poly(A) site cluster.

H: distance between the previous poly(A) site cluster region and the current poly(A) site cluster.

4   Notes

1. Starting with the first tract, *FindTail* then determines whether the downstream tracts can be merged to form a longer poly(A) fragment using an adjustable gap. After the merging step, it calculates the identity for all resultant poly(A) fragments and

remaining un-merged tracts, and filters them using an adjustable minimum identity (i.e., 85 % of A). Finally, the poly(A) length is calculated and filtered by an adjustable minimum length (i.e., 15 nt) required for a valid poly(A) tail. Thus, both perfect and imperfect poly(A) tails can be determined by *FindTail*.

2. In this step, users can further control the quality of poly(A) reads by applying more stringent criteria. Not only you can use *FilterIdentity* to extract sequence reads containing poly(A) tails with better quality (identity) (e.g., 90 % or above), but also you can control the minimum length of the cleaned sequence fragment (e.g., 25 nt) immediately before the poly(A) tails. This step can filter out a lot of sequence reads that will generate ambiguous mapping results in mRNA-to-genome mapping stage, so that we can save computational resources.

3. This is an optional step, depending on which mRNA-to-genome mapping software that you choose. We have compared with different mRNA-to-genome mapping software tools such as Helicosphere [2], GSNAP [5], and GMAP [6], and found that Helicos is able to provide better mapping results. So we decided to use Helicos in our pipeline. Unfortunately, Helicos cannot map any sequence read that contains the ambiguous nucleotide N. Since our reads have passed through the aforementioned quality control steps for both raw reads and poly(A) reads, we have obtained high-quality poly(A) reads that usually contain a few ambiguous nucleotides (e.g., 3 nt). So it is reasonable to conduct the nucleotide substitution.

4. Users may adjust the parameter "--read_limit" to apply more stringent criterion on the amount of supported sequence reads. The value of "--read_limit" depends on user's needs, but we suggest this value is over 3. After we obtain the high quality poly(A) sites, it is clear those poly(A) sites are often one to few nucleotide away due to the heterogeneity of poly(A) sites [7]. The high quality poly(A) sites display the whole polyadenylation landscape. The high quality poly(A) sites are basic foundation of poly(A) site clusters. The heterogeneity of poly(A) sites can be processed in cluster step (*see* Subheading 3.8).

5. Due to the sequencing orientation of the original data, we found two major outputs available from SRA data: poly(A) tail at 3′-end and poly(T) tail at the 5′-ends. These two situation are normal polyadenylation results. We suggest user to separate them into different processes, because A rich section and T rich section maybe appear at same sequence read.

6. These four outputs are annotated here: *pl* outputs are the sequences include "--ptype" homopolymer in left hand side of the sequences. *pr* outputs are the sequences which include "--ptype" homopolymer in right hand side of the sequences. *sl* outputs are the sequences which include "--stype" homopolymer in left hand side of the sequences. *sr* outputs are the sequences which include "--stype" homopolymer in right hand side of the sequences. For example, if users setup "--ptype" homopolymer is "A" and select the *pl* as output option, the output result will only maintain A rich tail at 5′end of sequence read, vice versa.

7. For detailed information about each parameter used in Helisphere toolkit, please refer to relevant document within the software download package. Here, we want to emphasize that we have utilized "--best_only" and "--min_score 4.3" to get the unique map results for each individual sequence read.

8. **CleanSam** program relies on the software GMAP for indexing and extracting genome sequences that is needed for filtering out internal priming poly(A) sites. Please read the GMAP document for installation and GMAP genome database setup. Once you finish GMPA installation and genome database setup, you can modify the aforementioned cleansam_pr_pa.pl and cleansam_sl_pt.pl based on the following examples:

```
my $gmapdir = "/usr/local/genome/gmap/gmap-
2012-12-12/bin/";
my $database='t3702.tair10.gmap20110831.k12';
```

To handling the internal priming, we retrieve the genomic sequence –10 to +5 nt surrounding the poly(A) site and examine. If this region has eight continuous As or more than eight As, it is considered as internal priming candidate. This read will be trimmed out from result.

9. We suggest the minimum value of "--read_limit" is 3, which means that a high-quality poly(A) site must have at least three high-quality poly(A) reads aligned unambiguously to the genomic location to support the genomic poly(A) site. If user want to apply more stringent criteria here, the maximum of this value is no limit. That depends on how many sequencing data has been processed. More data maybe need larger value at here.

Acknowledgement

This project was supported by a grant from the US National Institutes of Health (NIH-AREA) (1R15GM94732-1 A1 to CL and QQL), and by US National Science Foundation (grant nos. IOS–0817829 and IOS-1353354 to QQL).

Reference

1. Wheeler DL, Barrett T, Benson DA, Bryant SH, Canese K et al (2007) Database resources of the National Center for Biotechnology Information. Nucleic Acids Res 36:D13–D21, PMID: 18045790
2. Ozsolak F, Kapranov P, Foissac S, Kim SW, Fishilevich E, Monaghan AP, John B, Milos PM (2010) Comprehensive polyadenylation site maps in yeast and human reveal pervasive alternative polyadenylation. Cell 143:1018–1029, PMID: 21145465
3. Li H, Handsaker B, Wysoker A, Fennell T, Ruan J et al (2009) The Sequence alignment/map (SAM) format and SAMtools. Bioinformatics 25:2078–2079, PMID: 19505943
4. Tian B, Hu J, Zhang H, Lutz CS (2005) A large-scale analysis of mRNA polyadenylation of human and mouse genes. Nucleic Acids Res 33:201–212, PMID: 15647503
5. Wu TD, Nacu S (2010) Fast and SNP-tolerant detection of complex variants and splicing in short reads. Bioinformatics 26:873–881, PMID: 20147302
6. Wu TD, Watanabe CK (2005) GMAP: a genomic mapping and alignment program for mRNA and EST sequences. Bioinformatics 21:1859–1875, PMID: 15728110
7. Pauws E, van Kampen AH, van de Graaf SA, de Vijlder JJ, Ris-Stalpers C (2001) Heterogeneity in polyadenylation cleavage sites in mammalian mRNA sequences: implications for SAGE analysis. Nucleic Acids Res 29:1690–1694, PMCID: PMC31324

Chapter 4

Poly(A)-Tag Deep Sequencing Data Processing to Extract Poly(A) Sites

Xiaohui Wu, Guoli Ji, and Qingshun Quinn Li

Abstract

Polyadenylation [poly(A)] is an essential posttranscriptional processing step in the maturation of eukaryotic mRNA. The advent of next-generation sequencing (NGS) technology has offered feasible means to generate large-scale data and new opportunities for intensive study of polyadenylation, particularly deep sequencing of the transcriptome targeting the junction of 3′-UTR and the poly(A) tail of the transcript. To take advantage of this unprecedented amount of data, we present an automated workflow to identify polyadenylation sites by integrating NGS data cleaning, processing, mapping, normalizing, and clustering. In this pipeline, a series of Perl scripts are seamlessly integrated to iteratively map the single- or paired-end sequences to the reference genome. After mapping, the poly(A) tags (PATs) at the same genome coordinate are grouped into one cleavage site, and the internal priming artifacts removed. Then the ambiguous region is introduced to parse the genome annotation for cleavage site clustering. Finally, cleavage sites within a close range of 24 nucleotides and from different samples can be clustered into poly(A) clusters. This procedure could be used to identify thousands of reliable poly(A) clusters from millions of NGS sequences in different tissues or treatments.

Key words Polyadenylation site, Next-generation sequencing, Genomic data, Poly(A) clusters, Bioinformatic processing, PAT-seq

---

1 Introduction

Polyadenylation is a critical posttranscriptional processing step in the maturation of eukaryotic mRNA [1]. The location where the pre-mRNA is cleaved (also known as the poly(A) site) marks the end of a mRNA transcript. Many eukaryotic genes possess two or more poly(A) sites [2–4], and thus are involved in alternative polyadenylation (APA). APA is a powerful pathway that entails the selection of alternate poly(A) sites in a pre-mRNA and leads to the production of multiple mature mRNA isoforms from the same gene, resulting in potential gene expression regulation [1].

Arthur G. Hunt and Qingshun Quinn Li (eds.), *Polyadenylation in Plants: Methods and Protocols*, Methods in Molecular Biology, vol. 1255, DOI 10.1007/978-1-4939-2175-1_4, © Springer Science+Business Media New York 2015

While the significance of APA has been demonstrated in recent years, the scope and prevalence of it still remain to be further explored and understood for many genomes. High throughput next-generation sequencing (NGS) technologies have provided us the sequences in greater depth and coverage. Recent studies using NGS data has shown that over 70 % of Arabidopsis genes and over 80 % rice genes use APA sites [4, 5]. The coming flood of NGS data creates new opportunities for the more comprehensive study of the polyadenylated transcriptome, including several ways to just target the junction of 3′-UTR and the poly(A) tail [6]. Towards efficiently process these data, we have designed an automated workflow to map these poly(A) tags and identify poly(A) sites, by integrating NGS data cleaning, processing, mapping, normalizing, and clustering into one pipeline. This procedure is designed to be generic for different formats (e.g., Fasta, Fastq), or different types (e.g., single-end, paired-end of Illumina platform outputs) of NGS data. Thus, it could be adopted to use data from any organisms.

2   Materials

2.1   Equipment

Hardware: 64-bit computer running Linux.

Please make sure that Perl (http://www.perl.org/), MySQL (http://dev.mysql.com) and Bowtie (http://bowtie-bio.source-forge.net/index.shtml) are installed in your system.

2.2   Data

The NGS data could be downloaded from NCBI SRA, including two wild type leaf datasets from paired-end sequencing in Arabidopsis [4].

2.3   Setting Up

Download Perl scripts for this pipeline from http://www.users.miamioh.edu/liq/links.html; or http://www.polyA.org. Put these scripts in any directory.

3   Methods

3.1   Schema of the Poly(A) Site Identification Workflow

This poly(A) site identification pipeline integrates several steps. First, the annotation file in GFF format is parsed into the required format. Then single-end or paired-end sequences are mapped to the reference genome using an iterative mapping procedure. After mapping, poly(A) tags at the same coordinate are grouped into one cleavage site, and potential internal priming cleavage sites are removed. Next, cleavage sites close to each other in the same gene are clustered into poly(A) clusters (PACs). Finally, files storing cleavage sites and files recording PACs are imported into a PAC database for further analysis of polyadenylation or APA, such as

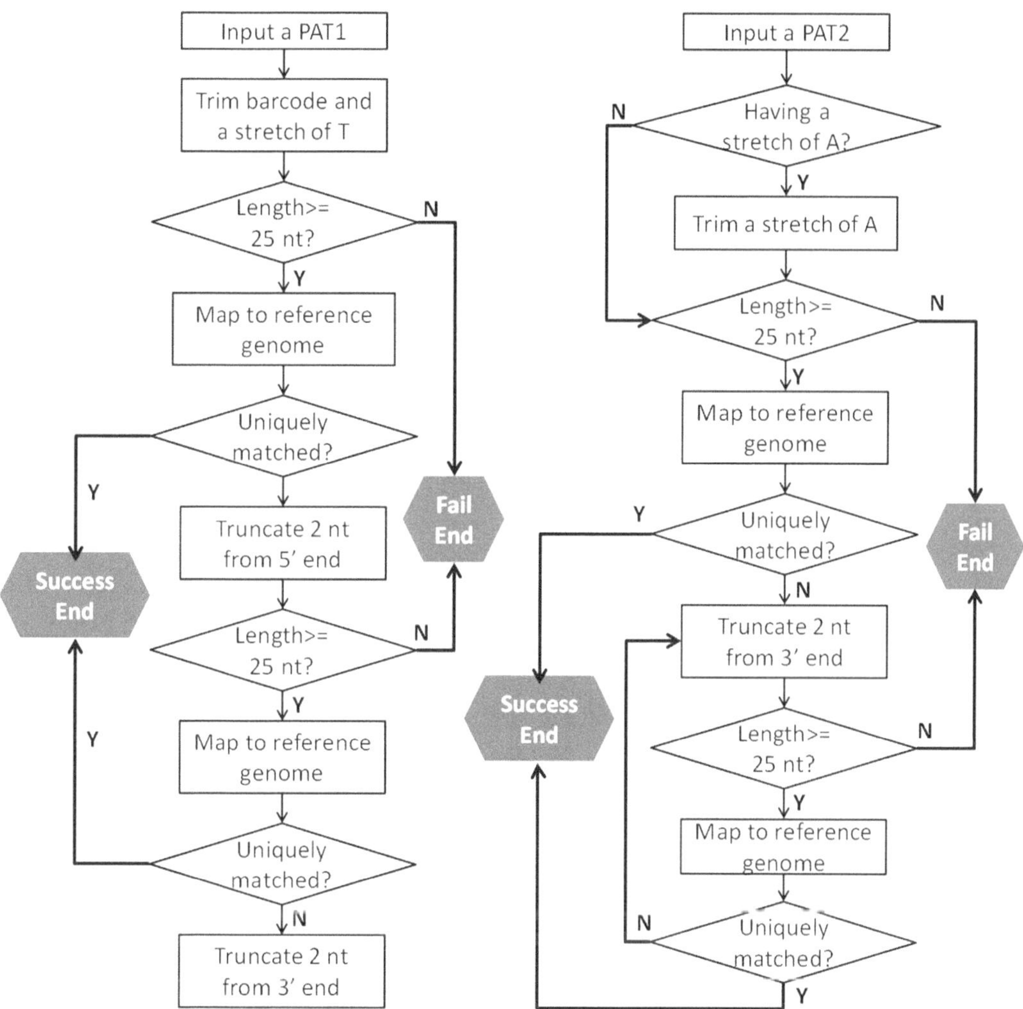

Fig. 1 PAT mapping for PAT1 and PAT2

detecting differentially expressed PACs and categorizing PACs based on their locations.

The flow chart of the iterative PAT mapping is shown in Fig. 1. For sequences from single-end sequencing (e.g., Read-1 from Illumina HiSeq 2000), the flow for PAT1 is sufficient for mapping. If the sequences are from paired-end sequencing, both the PAT1 and PAT2 mapping should be carried out to obtain candidate loci. In PAT1 mapping flow, PAT1 with the terminal barcode and a run of eight or more consecutive Ts are identified and the sub-sequence after the oligo(T) is also trimmed. Sequences with length shorter than 25 nt after trimming are discarded. Otherwise, it is remained for further mapping. Since a variable length of adaptor is expected in the remaining sub-sequence, an iterative process is implemented which is designated to minimize the impact of non-templated nucleotide addition. Bowtie [7], a short read aligner, is used in our pipeline for short sequence mapping. Only unique hit and at most two nt mismatches are allowed for a perfect match. If there is no perfect match for this sequence, then its 3′-most two nts are trimmed for mapping again. This process is repeated until the sequence is shorter than 25 nt or multiple hits are found.

If the raw data is from paired-end sequencing, the paired-end partners (PAT2) to the mapped PAT1 are then filtered and divided into two groups: PAT2_A and PAT2_N. PAT2_A stores the sequences with at least 8 consecutive As (i.e., poly(A) tail). PAT2_N contains sequences without such poly(A) stretch. The sequences in PAT2_A are trimmed off the poly(A) tail. Both sequences in PAT2_A and PAT2_N are then mapped to the genome using the same iterative process as PAT1 mapping to recover as many authentic PATs as possible.

After PAT mapping, the coordinates of the mapped sequences are recorded and related with genome annotation to define their genomic locations. If the sequences are not from paired-end sequencing, then the coordinates of PAT1 are the final candidate cleavage sites. If they are from paired-end sequencing, then the candidate cleavage sites are determined using the following criteria: both PAT1 and PAT2 need to map to the same gene, or to the same chromosome on the same strand (either + or −) and within a distance less than 1,000 nt. After mapping, the candidate cleavage sites that represent possible internal priming by reverse transcriptase are discarded.

Figure 2 shows some possible cases of paired-end sequences. If PAT2 is with a poly(A) tail (type PAT2_A), the oligo(A) of the sequence in PAT2 is trimmed, so as the oligo(T) of the PAT1 trimmed. Both sequences are mapped to the same locus, and the cleavage site can be defined as marked (Fig. 2a). If PAT2 is without a poly(A) tail (type PAT2_N), then PAT2 and PAT1 will be mapped to different loci but within a certain distance (Fig. 2b).

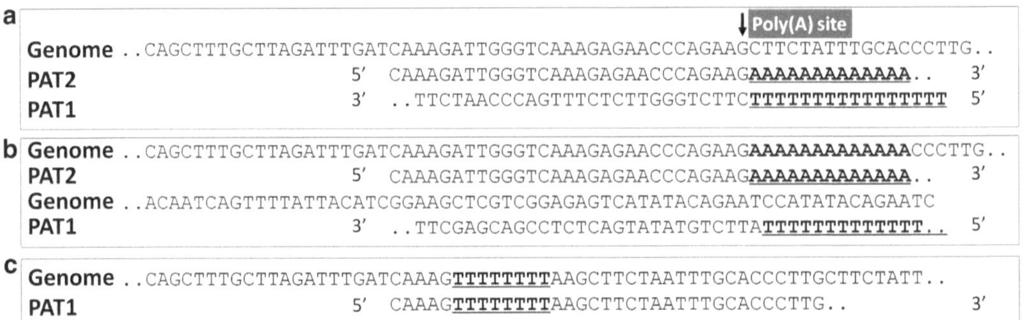

Fig. 2 Example cases of paired-end sequences. (**a**) PAT1 has a poly(A) tail, and PAT1 and PAT2 map to the same locus; (**b**) PAT1 and PAT2 map to different loci; (**c**) A case of internal priming

Interestingly, there is a stretch of A in the PAT2 in Fig. 2b, while this A stretch is not a real poly(A) tail of mRNA but from the reference genome. This result also indicates that our mapping procedure has the ability to differentiate the real poly(A) tail from the mere A stretch. Figure 2c shows a case of internal priming. Though the PAT1 can be mapped to the genome, there is a stretch of T (or the complementary As) around the candidate cleavage site. This T stretch is actually from the reference genome rather than the mRNA, and therefore this cleavage site is not real and is discarded.

3.2 Implementation of the Poly(A) Site Identification Flow

This pipeline was implemented by several Perl scripts, integrating some in-house tools as add-ons.

1. Parsing: UTIL_parseTair.pl is a Perl script for parsing the annotation file of Arabidopsis in GFF format into the required format.

2. For the iterative mapping procedure, the following scripts were used:

```
MAP_filterPolySeq.pl: filtering the raw tags
starting with T-stretch or ending with A-stretch.
    MAP_divide12tag.pl: separating the paired
end tags into different groups for paired end
mapping.
    MAP_polySeq2PA.pl: mapping the PATs by Bowtie
to get the coordinates for the mapped PATs.
    MAP_JoinTAN.pl: joining the mapped paired
end tags to make sure the mapped poly(T) tags
could be paired by the tags within the same
gene or within a given distance.
    MAP_parseBwt2PNP.pl: parsing the Bowtie
mapping results to the coordinates of the
mapped tags.
```

3. Grouping: FILE_PAT2PA.pl is used to group the poly(A) tags (PATs) at the same coordinate into one cleavage sites.

4. Removal of internal priming: PAT_setIP.pl is used to remove the internal priming cleavage sites.

5. Clustering: PAT_PA2PAC.pl is used to cluster the cleavage sites into poly(A) clusters (PACs).

6. PAC database: two Perl scripts PAT_alterPA.pl are used to parse the files storing the cleavage sites into the PAC database (*see* **Note 1**).

3.3 Install the Genome Sequence Database for Bowtie

1. Run the following command in the Linux/Unix console to install the genome sequence database:

```
bowtie-build genome_sequence.fasta dbname
mv dbname* path_of_bowtie/indexes/
cp   genome_sequence.fasta   path_of_bowtie/
indexes/ dbname.fa
```

2. Modify .bashrc file to add the database indexes to the system path:

```
export BOWTIE_INDEXES=path_of_bowtie/indexes/
```

3. Test whether Bowtie and the genome sequence database are successfully installed:

```
bowtie -c dbname ATGAACCCTGTCGACCATCCCCATGG
```

3.4 PAT Mapping for Single-End Sequences

For sequences from single-end sequencing (*see* **Note 2**), using MAP_polySeq2PA.pl to do the mapping. This script mainly contains the following options:

seq: sequence file in fataq or fasta format.

bwtopt: Bowtie mapping option, default is "-v 2 –m 1".

bwtdb: index of Bowtie genome sequence database.

poly: T/A; to set the input sequence starts with poly(T) or ends with poly(A).

fml: min length after filtering the raw sequences.

tml: min length for the iterative PAT mapping.

t5/t3: t5 = 2 × 2 means trim 2 nt each time from the 5′ end for twice until the length of the sequence is shorter than tml.

mode: can be 53/35/5353/3535 to set the order of t3 and t5.

You can use the following command to map single-end sequence file:

```
MAP_polySeq2PA.pl -seq "fasta_or_fastq_seq_
file" -bwtdb dbname -poly T -fml 25 -tml 25
-t5 2X1 -t3 2XN -mode 53
```

You may change the Bowtie option to map the fastq file and consider the quality scores:

```
MAP_polySeq2PA.pl -seq "fasta_or_fastq_seq_
file" -bwtopt " -n 1 -l 25 -m 1 -e 70--est--
trata" -bwtdb dbname -poly T -fml 25 -tml 25
-t5 2X1 -t3 2XN -mode 53
```

3.5 PAT Mapping for Paired-End Sequences

For sequences from paired-end sequencing (*see* **Note 2**), use MAP_RUN_PairMap.pl to do the mapping. This script mainly contains the following options:

tag1/tag2: sequence file in fataq or fasta format for poly(A) sequence (tag1) or poly(T) sequence (tag2).

bwtopt: Bowtie mapping option, default is "-v 2 –m 1".

bwtdb: index of Bowtie genome sequence database.

poly: T/A; to set the input sequence starts with poly(T) or ends with poly(A).

You can use the following command to map paired-end sequence file:

```
MAP_RUN_PairMap.pl -tag1 polyA_seq.fq -tag2
polyT_seq.fq -bwtdb dbname
```

You may change the Bowtie option to map the fastq file and consider the quality scores:

```
MAP_RUN_PairMap.pl -bwtopt " -n 1 -l 25 -m 1 -e
70--est--trata" -tag1 polyA_seq.fq -tag2 polyT_
seq.fq -bwtdb dbname
```

3.6   Determination of Cleavage Sites

After PAT mapping, the coordinates of the mapped sequences are recorded. If the sequences are not from paired-end sequencing, then the PATs from the same coordinate are grouped into one candidate cleavage site. If they are from paired-end sequencing, then the candidate cleavage sites are determined using the following criteria: both sequences from the same sequencing cluster (from Illumina output) need to map to the same gene, or to the same chromosome on the same strands (either + or –) and within a distance less than 1,000 nt (*see* **Note 3**).

3.6.1. For single-end sequences, you can use the following command to group PATs from the same coordinate:

```
FILE_PAT2PA.pl -pat mapped_PAT_file -pa
output_PA_file -tcols 1:2:6
```

This script mainly contains the following options:

pat: the mapping result file from MAP_polySeq2PA.pl.

pa: output cleavage site file, containing four columns: chr, strand, coord, and tagnum.

tcols: user defined columns for chr, strand, and coord in the "pat" file, default is 1:2:6.

3.6.2. For paired-end sequences, you need to run FILE_TAN2PAT. pl to relate the poly(A) and poly(T) sequences first and then use FILE_PAT2PA.pl to group the sequences into cleavage sites.

```
FILE_TAN2PAT.pl -tan mapped_PAT_file -gene
all_genes_from_gff -ext 500 -dist 1000 -ofile
output_file
```

This script mainly contains the following options:

tan: the mapping result file from MAP_RUN_PairMap.pl.

gene: file recording the genes with their start and end coordinates, including columns: chr, strand, gene, start, end.

ext: extend gene for ext nts.

dist: min distance allowed for poly(A) and poly(T) sequence when they are not in the same gene.

Then you can use FILE_PAT2PA.pl with the same command as above to group PATs from the same coordinate.

3.7 Removal of Possible Internal Priming Candidates

After mapping, the candidate cleavage sites that represented possible internal priming by reverse transcriptase are discarded using a strategy similar to that described in [3]. You can use the following script to remove internal priming candidates (*see* **Note 4**):

```
PAT_setIP.pl -itbl input_file -otbl output_file
-iptbl output_internal_priming_file -flds 0:1:2
-format file -conf database_config.xml
```

This script mainly contains the following options:

itbl: the input file, with at least three columns: chr, strand, coord.

otbl: output file without internal priming tags.

iptbl: output file with internal priming tags.

flds: specify the columns for chr, strand, and coord.

format: table(default) or file.

3.8 Clustering of Micro-Heterogenic Cleavage Sites

It is suggested that most clusters are separated by 24 or more nucleotides within a transcription unit [2, 4, 8]; thus, a 24-nt interval was adopted to differentiate macroheterogeneity of poly(A) sites from microheterogeneity of poly(A) sites within the same transcript. The poly(A) site in a cluster supported by the greatest number of PATs was set as the reference site for the poly(A) site cluster (termed here as "PAC"), and the number of PATs associated with a PAC was defined as the sum of all tags supporting its constituent poly(A) sites.

You can use the following script to cluster cleavage sites into PACs:

```
PAT_PA2PAC.pl -d 24 -mtbl cleavage_site_table
-gfftbl genome_annotation_table -otbl output_
table -smps sample_column -conf database_con-
fig.xml
```

This script mainly contains the following options:

mtbl: cleavage site table with columns: chr, strand, coord, tot_tagnum, sample_column.

gfftbl: genome annotation table parsed from the gff format file.

otbl: output PAC table.

smps: sample column in "mtbl" to be grouped into PACs.

conf: the path of the XML file storing the configuration of MySQL database.

4   Notes

1. Based on the PAC database, we could make further analysis of polyadenylation or APA, such as detecting differentially expressed PACs and categorizing PACs based on their locations.

2. Single-end and paired-end sequences are the outputs from the Illumina sequencing platform. For regular RNA-seq data, only a small part of the raw sequences may be with A/T-stretch. Nevertheless, this pipeline could also be applied without modification to the RNA-seq data.

3. The number (1,000 nt) is arbitrary given the current sequencing capacity of Illumina platform may not beyond this length.

4. To avoid possible internal priming, the genomic sequences around the mapped poly(A) sites (i.e., −10 to +10 nt) are scanned for continuous adenines of more than 5 nt or more than 6 nt adenines in any 10 nt window. The PAT mapped to these sites is considered as internal priming candidate and will be removed.

Acknowledgement

Funding supports for this work were from the National Natural Science Foundation of China (Nos. 61174161 and 61304141), the Natural Science Foundation of Fujian Province of China (No. 2012J01154), the specialized Research Fund for the Doctoral Program of Higher Education of China (Nos. 20130121130004 and 20120121120038), and the Fundamental Research Funds for the Central Universities in China (Xiamen University: No. 2013121025), Xiamen Shuangbai Talent Plan (to QQL), and US National Science Foundation (grant nos. IOS–0817829 and IOS-1353354 to QQL).

References

1. Xing D, Li QQ (2011) Alternative polyadenylation and gene expression regulation in plants. Wiley Interdiscip Rev RNA 2(3):445–458. doi:10.1002/wrna.59

2. Shen Y, Ji G, Haas BJ, Wu X, Zheng J, Reese GJ, Li QQ (2008) Genome level analysis of rice mRNA 3′-end processing signals and alternative polyadenylation. Nucleic Acids Res 36(9): 3150–3161

3. Tian B, Hu J, Zhang HB, Lutz CS (2005) A large-scale analysis of mRNA polyadenylation of human and mouse genes. Nucleic Acids Res 33(1):201–212. doi:10.1093/nar/gki158

4. Wu X, Liu M, Downie B, Liang C, Ji G, Li QQ, Hunt AG (2011) Genome-wide landscape of polyadenylation in Arabidopsis provides evidence for extensive alternative polyadenylation. Proc Natl Acad Sci U S A 108(30):12533–12538. doi:10.1073/pnas.1019732108

5. Shen Y, Venu RC, Nobuta K, Wu X, Notibala V, Demirci C, Meyers BC, Wang G-L, Ji G, Li QQ (2011) Transcriptome dynamics

through alternative polyadenylation in developmental and environmental responses in plants revealed by deep sequencing. Genome Res 21(9):1478–1486. doi:10.1101/gr.114744.110

6. Ma L, Pati PK, Liu M, Li QQ, Hunt AG (2014) High throughput characterizations of poly(A) site choice in plants. Methods 67(1):74–83. doi:10.1016/j.ymeth.2013.06.037

7. Langmead B, Trapnell C, Pop M, Salzberg SL (2009) Ultrafast and memory-efficient alignment of short DNA sequences to the human genome. Genome Biol 10(3):R25. doi:10.1186/gb-2009-10-3-r25

8. Shen Y, Liu Y, Liu L, Liang C, Li QQ (2008) Unique features of nuclear mRNA poly(A) signals and alternative polyadenylation in Chlamydomonas reinhardtii. Genetics 179(1):167–176

Chapter 5

Analysis of Poly(A) Site Choice Using a Java-Based Clustering Algorithm

Patrick E. Thomas

Abstract

Modern high-throughput DNA sequencing has the potential to generate large volumes of data for analysis by investigators—including poly(A) site data. Here I describe a computational method to compare poly(A) site choice differences between two large data sets based on the relative abundance and position of tags within each reference sequence to which they are aligned. This method provides rapid quantification and visualization of differences and similarities in poly(A) site choice between the two datasets.

Key words Alternative polyadenylation, Java, Next generation sequencing, PAT-seq

1 Introduction

In higher eukaryotes, alternative polyadenylation is an important regulator of gene expression [1–5]. Global analysis of polyadenylation differences between samples is an avenue of research made possible by the availability of complete genome sequences and high-throughput sequencing technology and has been applied to the study of polyadenylation in plants [6–8]. The creation of poly(A) site aligned tags (PATs) as described by Wu and colleagues [8] can provide an investigator with high quality datasets designating poly(A) sites. Here we describe a simple data pipeline for producing meaningful and useful data from comparisons of large PAT datasets. This pipeline consists of any alignment software capable of generating SAM files showing tag alignments, an in-house authored Java-based program titled the Poly(A)Tag Alignment Profiling Program (PATAPP), and any spreadsheet program, such as Microsoft Excel, capable of reading tab-delimitated text files.

Using this pipeline we can generate summary files that show a reference sequence-by-reference sequence difference metric between the compared PAT sets. The difference metric can be calculated position-by-position or based on clustering in order to compensate

Arthur G. Hunt and Qingshun Quinn Li (eds.), *Polyadenylation in Plants: Methods and Protocols*, Methods in Molecular Biology, vol. 1255, DOI 10.1007/978-1-4939-2175-1_5, © Springer Science+Business Media New York 2015

for microheterogeneity. While this metric alone is very useful, we also describe various ways to visually show a comparison between samples using these files. Assuming the preferred clustering methodology is used, a file containing information on abundance of PATs in every cluster defined for each reference sequence can be also generated. Finally, our pipeline also has the capability to generate a file showing the position-by-position abundance of PATs within each reference sequence, which can quickly be made into a scatter plot to visualize the distributions. To minimize the numbers of files generated, the position-by-position files can be filtered based on their difference metric.

2 Materials

2.1 System Requirements

The systems that can be used for this analysis vary greatly, but will mostly depend upon which alignment software and spread sheet software the investigator chooses. PATAPP, which we provide for download, can be launched from any operating system that supports Java 7. It has been tested and found to function properly on Microsoft Windows XP and 7, Macintosh OS 10.6.x and later versions, and Ubuntu 12 Linux with no functional problems (*see* **Note 1**). PATAPP can successfully process files in the 2–4 GB range with 2 G of memory available and allocated to the Java virtual machine, but more memory may be needed for larger files.

2.2 Alignment Software

The tag generation and sequencing procedure generates a high number of sequence reads. With good source RNA and bench technique, the generation of these tags is accurate, efficient, and reproducible [9]. However, there will usually be tags generated that align to low complexity sequences and some limited internal priming artifacts. We highly recommend using an alignment program or series of programs to clean up tag data prior to or during alignment. For our analysis and generation we used the CLC Genomics Workbench. Alternative open-source programs that generate SAM files are suitable. The individual investigator should make choices appropriate to the needs of their investigation.

2.3 PATAPP Packages

PATAPP is a Java-based program with a GUI for ease of use on multiple platforms (*see* **Note 2**). The PATAPP program, test files, and other resources can be downloaded as an archive file (*see* **Note 3**). The PATAPP is an executable JAR file. The test file folder contains sample files for test runs and the results as they should be output so that you can verify the program is functioning correctly on your operating system.

2.4 Spreadsheet Software

PATAPP generates files in tab delimited text format. These files are viewable in a text viewer and useful data can be extracted from them in that format, but it is not ideal. The investigator will be much better served by opening these files in any spreadsheet software (such as Microsoft Excel) which will open tab delimited text. The text files will be organized in such a way as to make visualization of results very simple by methods we describe if spreadsheet software is used.

3   Methods

3.1 Sequence Read Alignment

1. The tag sets you wish to compare should be in two separate batches of sequence data. Tag sequences must be trimmed to remove linkers, bar codes, sequencing adapters/primers, and oligo-dT tracts before mapping (*see* **Note 4**).

2. Using your desired alignment software, map each set of tags to your reference sequences (*see* **Note 5**). Parameters should be chosen so that low complexity sequences do not map. Additionally, genomic positions that are adjacent to runs of 6 or more "A" nucleotides should be masked during mapping. This helps to remove any possible internal priming artifacts and false matches.

3. The data from the mappings should be exported into individual SAM files, one for each PAT data sample.

3.2 Use of PATAPP

1. Download the PATAPP zip file.

2. Unzip the PATAPP file.

3. Launch the PATAPP using one of the following options:

   (a) In the command line prompt of your operating system type, navigate to the folder containing the PATAPP download. Type "Java -jar PATAPP.jar", and hit enter. You may add additional arguments to this line to control the memory of the virtual machine or other features of the virtual machine (*see* **Note 6**).

   (b) On most systems, double click on the PATAPP.jar file. This will open the PATAPP with the default Java virtual machine argument. This configuration will be undesirable for running large files.

4. Basic directions for running PATAPP are contained in the display window on the left side and should be reviewed before using the program.

5. Choose the FASTA formatted file that contains the reference sequences you mapped your tag sets to by clicking the "browse files" button beside the label "Choose the Fasta file to which your tags are aligned". Once you click this button a file chooser panel will appear. Navigate to the location of your input file, select it, and click open. The path to the file should now be displayed in the text area next to "browse files" button.

6. Use the same basic procedure as outlined in **step 5** to choose both the .SAM files containing your alignment data by clicking the "browse files" buttons next to the corresponding labels. It should be noted that from this point on Sam File 1 will refer to the file selected in the first text area, and that Sam File 2 will refer to the file selected in the second text area (*see* **Note 7**).

7. Choose the directory to which you want the output saved (*see* **Note 8**). Selecting this works similar to choosing the files in **steps 5** and **6**, but now the file chooser will only let you select directories (*see* **Note 9**).

8. Select the criteria for the analysis by entering numbers in the text fields on the bottom left. Default values are automatically generated if you wish to accept them. The different parameters you may manipulate are:

    (a) Minimum Tags per Reference Sequence (MTRS): This sets the minimum number of tags that must have aligned to a reference sequence *in both sets* before a reference's difference score will be reported. For instance, if MTRS is set to 20, and 1,000 tags aligned to reference sequence A in the first SAM file, but only 19 tags aligned to that same sequence in the second SAM file no data would be reported in the final report for sequence A. MTRS is set to 15 by default.

    (b) Minimum Percentage of Parent Per Cluster (MPPPC): This is a filter to remove clusters which may have very low tag abundance and are thus potentially artifacts or of negligible impact. For instance, if the MPPPC was set to 3, it would require all clusters formed to contain at least 3 % of the total number of tags aligned to the sequence in order to be used in the analysis. In this example if a reference sequence had 3,000 tags aligned from one set, in order for a cluster to be considered significant it would have to include at least 90 tags. MPPPC is set to 0 by default.

    (c) Maximum Distance Between Tags (MDBT): This controls the gap allowed between tags for them to be considered in the same cluster. A setting of 1 would require alignments to be next to each other with no gaps in order to be clustered together. MDBT is set to 10 by default.

    (d) Maximum Cluster Size: This limits the amount that clusters can grow before tags will no longer be grouped together. Clustering helps to compensate for the micro-heterogeneity that is ubiquitous in plant poly(A) site choice. Decreasing this number to 1 will cause each tag to be calculated as its own cluster. Choosing to large a number makes the analysis meaningless by causing the difference to metric approach 0 for all sequences. MCS is set to 30 by default.

9. Click the check boxes to choose the type of output files desired. The options are:

    (a) Standard comparison file: This file contains a list of every reference sequence that was compared between the two sets. It reports the name of the sequence, the length of the reference sequence, the number of tags aligned in each SAM file, and the cluster comparison metric. The metric is on a scale of 0–1, with 0 meaning identical poly(A) site choice and 1 meaning totally different poly(A) site choice.

    (b) SAM 1 and SAM 2 individual site files: These will print the specifics of the clusters created by PATAPP for each reference in each file. By default no sequence is printed with this option, but the user may select to have the sequence for each cluster printed as a string of characters in one cell or have each nucleotide separated by a tab character. The sequence is printed based on the point in the cluster with the highest tag abundance. The user may specify how many bases upstream and downstream of the sequence that the program should output.

    (c) Single Reference Image Files: This option produces a file *for each reference sequence* which was compared. These files contain two rows representing each position along the reference sequence which indicate the number of tags that aligned at each position. In the case of a large data set this could be thousands of files; so it is recommended that the user selects a range of difference metrics to filter the output by. For instance, to only see sequences with very large difference the user could choose to print from 0.7 to 1.

10. After all options are selected, verify that they are correct, and click the execute button.

11. The analysis could take several minutes for large files. When it is complete a message box will pop up to notify you that the process is complete. Click OK in the notification box. Your output files are now in the specified output directory.

3.3 Creation of a Running Sum Plot

Running sum plots allow for visualization of differences between samples. The running sum plot requires that the investigator generated a standard comparison file using PATAPP. There are several ways to create the running sum plot and techniques will vary based on the chosen spreadsheet software. This procedure describes the use of the "Running Sum Template.xlsx" file included with the PATAPP download.

1. Open the standard comparison file from your PATAPP analysis using your spread sheet software.

2. Open the "Running Sum Template" file.

3. Either copy the data from your standard comparison file into the template in the designated areas, or copy columns G–J from the template into your file.

4. Column "G" is the column that rounds data to three digits so that it can be counted properly by Excel. You should copy and paste the function in column "G" so that every metric is rounded. If you have 1,000 reference sequences this would mean you would copy and paste the round function down to line 1001.

5. Column "I" is where the counting of each possible metric value is performed and converted to a percentage. In order for the functions to work properly the range must be adjusted. In the cell "I2" of the template, the function reads "=COUNTIF(G\$2:G\$10,H2)/COUNT(G\$2:G\$10)". Both occurrences of the number 10 should be replaced with the correct ending range of data in your file. If there were 1,000 reference sequences the formula would look like "=COUNTIF(G\$2:G\$1001,H2)/COUNT(G\$2:G\$1001)".

6. Copy cell "I2" and paste the functions into cells "I3" through "I1002".

7. Select cells "H2" through "H1002" and "I2" through "I1002", and on the insert tab's scatter plot drop-down menu select "scatter with smooth lines".

---

4 Notes

1. On Ubuntu authentic Oracle Java 7 should be used, and not the open source version of Java that is made available on many Ubuntu systems.

2. The GUI has some resizing issues on Ubuntu Linux and some of the tested Mac operating systems. The program is still fully functional, but the window occasionally has to be moved or manually resized in order access all parts of the program.

3. To obtain the archive file for installing PATAPP, email Dr. Arthur G. Hunt (aghunt00@uky.edu). In the email, indicate the desire to obtain the program, and include full contact information. This will be used to monitor interest and to maintain a database of users who may be informed whenever the program is updated.

4. PATAPP compares the locations of the first nucleotides (reading from left-to-right along a reference sequence) to which individual tags are mapped. To avoid instances where the mapping program includes mismatched ends in the definitions of mapped locations in the .SAM files, it is necessary to completely

trim tag sequences so that only mRNA body-derived sequences are present.

5. The reference sequence database can be any collection of sequences. Because PATAPP compares the locations of the first nucleotide (reading left-to-right along a reference sequence) to which individual tags are mapped, and since poly(A) sites lie at the 3′ ends of the corresponding transcripts, the orientation of the reference sequences should be inverted such that the leftwards most sequence (reading left-to-right along the sequence) corresponds to the 3′-most position of the reference sequence; this is accomplished by reversing and complementing the desired reference sequence database. In addition, PATAPP only reads tags that align in the orientation suited for analysis of "sense" poly(A) sites (e.g., those sites that correspond to sites in the mRNA sense) and will filter out the rest. If antisense poly(A) sites are to be analyzed, the orientation of the reference sequences should be inverted (e.g., the original "sense" orientation should be used). We have used an extended 3′-UTR reference database for *Arabidopsis*, but the references may consist of full genes, other focused parts of genomes and transcripts (e.g., introns, coding regions, etc.), or entire chromosomes. The choice of reference database will determine the nature and size of the eventual output file. For example, in the cited examples [6, 9], differences in 3′-UTR-localized poly(A) sites were assessed. By using complete genes, a gene-by-gene evaluation of changes in poly(A) site usage throughout the transcription unit is possible.

6. Included in the PATAPP archive is a text file which contains a recommended launch configuration command which would launch PATAPP with 1.6 G of memory. This is enough memory to complete most of the analyses we have run. The name of the file is "LaunchCommand.txt".

7. The results of the comparison are not affected by which file is selected as "Sam File 1" or "Sam File 2", but it can have an effect on the speed of program execution. If your alignment files are of drastically different sizes, choosing the smaller file as "Sam File 1" will speed the filtering process and reduce memory usage.

8. Files generated by PATAPP always have the same general names; so, if you wish to do multiple concurrent runs, you must select a different output directory every time or your files will be overwritten without warning by each subsequent run. You may manually rename files between runs and then keep the same output directory each time.

9. You may create a new directory from the file chooser and then select that directory.

Acknowledgements

This work was supported by the National Science Foundation (award MCB-0313472 to Drs. Arthur G. Hunt and Q. Quinn Li, and an RET supplement to award IOS-0817818).

References

1. Mueller AA, Cheung TH, Rando TA (2013) All's well that ends well: alternative polyadenylation and its implications for stem cell biology. Curr Opin Cell Biol 25(2):222–232. doi:10.1016/j.ceb.2012.12.008

2. Elkon R, Ugalde AP, Agami R (2013) Alternative cleavage and polyadenylation: extent, regulation and function. Nat Rev Genet 14(7):496–506. doi:10.1038/nrg3482

3. Shi Y (2012) Alternative polyadenylation: new insights from global analyses. RNA 18(12):2105–2117. doi:10.1261/rna.035899.112

4. Xing D, Li QQ (2011) Alternative polyadenylation and gene expression regulation in plants. Wiley Interdiscip Rev RNA 2(3):445–458. doi:10.1002/wrna.59

5. Hunt AG (2008) Messenger RNA 3′ end formation in plants. Curr Top Microbiol Immunol 326:151–177

6. Thomas PE, Wu X, Liu M, Gaffney B, Ji G, Li QQ, Hunt AG (2012) Genome-wide control of polyadenylation site choice by CPSF30 in Arabidopsis. Plant Cell 24(11):4376–4388. doi:10.1105/tpc.112.096107

7. Sherstnev A, Duc C, Cole C, Zacharaki V, Hornyik C, Ozsolak F, Milos PM, Barton GJ, Simpson GG (2012) Direct sequencing of Arabidopsis thaliana RNA reveals patterns of cleavage and polyadenylation. Nat Struct Mol Biol 19(8):845–852. doi:10.1038/nsmb.2345

8. Wu X, Liu M, Downie B, Liang C, Ji G, Li QQ, Hunt AG (2011) Genome-wide landscape of polyadenylation in Arabidopsis provides evidence for extensive alternative polyadenylation. Proc Natl Acad Sci U S A 108(30):12533–12538. doi:10.1073/pnas.1019732108

9. Ma L, Pati PK, Liu M, Li QQ, Hunt AG (2013) High throughput characterizations of poly(A) site choice in plants. Methods 67(1):74–83. doi:10.1016/j.ymeth.2013.06.037

Chapter 6

RADPRE: A Computational Program for Identification of Differential mRNA Processing Including Alternative Polyadenylation

Denghui Xing and Qingshun Quinn Li

Abstract

Genome-wide studies revealed the prevalence of multiple transcripts resulting from alternative polyadenylation (APA) of a single given gene in higher eukaryotes. Several studies in the past few years attempted to address how those APA events are regulated and what the biological consequences of those regulations are. Common to these efforts is the comparison of unbiased transcriptome data, either derived from whole-genome tiling array or next generation sequencing, to identify the specific APA events in a given condition. RADPRE (*Ratio*-based *Analysis* of *Differential* mRNA *Processing* and *Expression*) is an R program, developed to serve such a purpose using data from the whole-genome tilling array. RADPRE took a set of tilling array data as input, performed a series of calculation including a correction of the probe affinity variation, a hierarchy of statistical tests and an estimation of the false discovery rate (FDR) of the differentially processed genes (DPG). The result was an output of a few tabular files including DPG and their corresponding FDR. This chapter is written for scientists with limited programming experiences.

Key words Alternative polyadenylation, Posttranscriptional processing, RADPRE, Tiling array

1 Introduction

The alternative processing of a pre-mRNA could result in multiple transcripts from a single given gene and therefore is an important way to increase the coding capacity of a genome with a limited number of genes. The past years have witnessed a dramatic and continuous increase of a number of alternative processing events based on the genome-wide studies of a variety of eukaryotic organisms [1–3]. The challenge now is on how the alternative pre-mRNA processing including APA is regulated and what the biological consequences of those regulations are. Toward this end, the very first step is to identify the specific differential processing events in a given condition, be it a mutant, a developmental stage or a tissue type. One approach to address this question is to

Arthur G. Hunt and Qingshun Quinn Li (eds.), *Polyadenylation in Plants: Methods and Protocols,* Methods in Molecular Biology, vol. 1255, DOI 10.1007/978-1-4939-2175-1_6, © Springer Science+Business Media New York 2015

compare transcriptomes derived from two or multiple conditions. Critical to this approach is the unbiased sampling of the transcriptomes. Both whole-genome tiling array and next generation sequencing meet this requirement. For methods using sequencing approach, the "3P-Seq" and "PAT-Seq", which were specifically dealing with alternative polyadenylation, are recommended [1, 4].

RADPRE was designed to use the transcriptome data from tilling arrays [5]. The underlining principle of how the tiling data might be employed to identify the differential APA events was illustrated in Fig. 1 and detailed previously [5]. The RADPRE pipeline constituted three major steps. In the first step, RADPRE took a set of tiling array data as input and calculated the ratios of normalized probe intensities between two conditions (Fig. 2a).

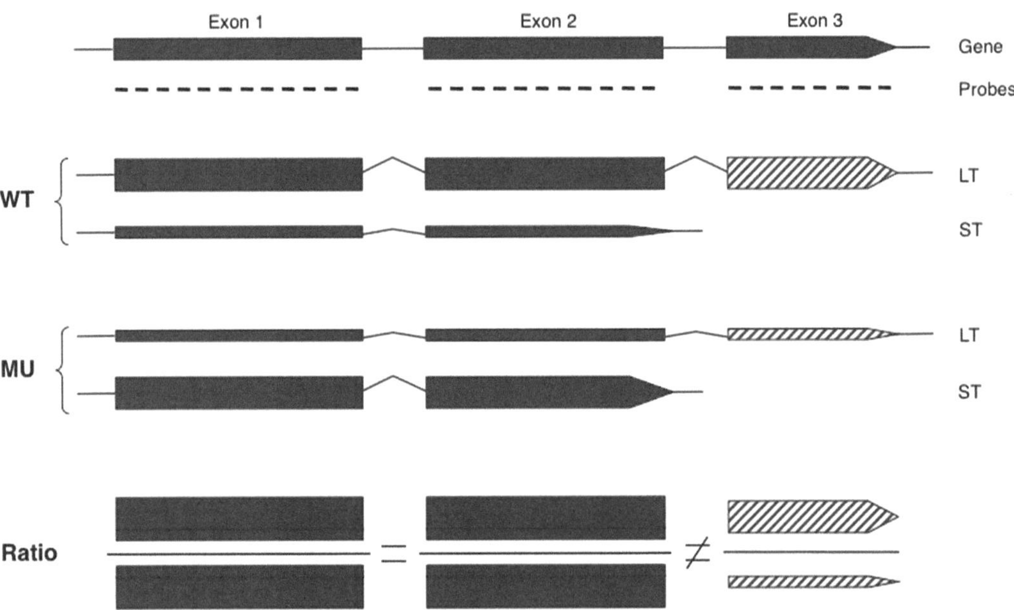

Fig. 1 Identification of DPG genes based on the ratios of its exons between two states (wild type WT and mutant MU). The gene structure was shown on the top of the graph with *filled boxes* denoting exons, *lines* denoting introns, and the *short lines* under the exons denoting the tiling array probes. The gene could generate two transcripts, a long (LT) and a short (ST), with LT derived from a distal poly(A) site and containing exon 3 (*hatched box*), and ST from a proximal poly(A) site within intron 2. The thickness of the box represents the relative abundance of the transcripts. The relative abundance of two transcripts was altered between WT and MU due to the shift of the poly(A) site usage between WT and MU. The measured abundance of each exon was based on its corresponding probes, which reflected the sum of two transcripts. The measured abundance of exon 1 and 2 was the same between WT and MU, but that of exon 3 was different. Therefore, the ratio of exon 1 and 2 between WT and MU was equal to 1, but different from the ratio of exon 3. In that case, the poly(A) site choice was not affected by the mutant and the relative abundance of the two transcripts would be the same between WT and MU. The ratio of the measured abundance between WT and MU would be equal for all three exons. Reused from ref. 5 as permitted by the copyright statement in the original publication

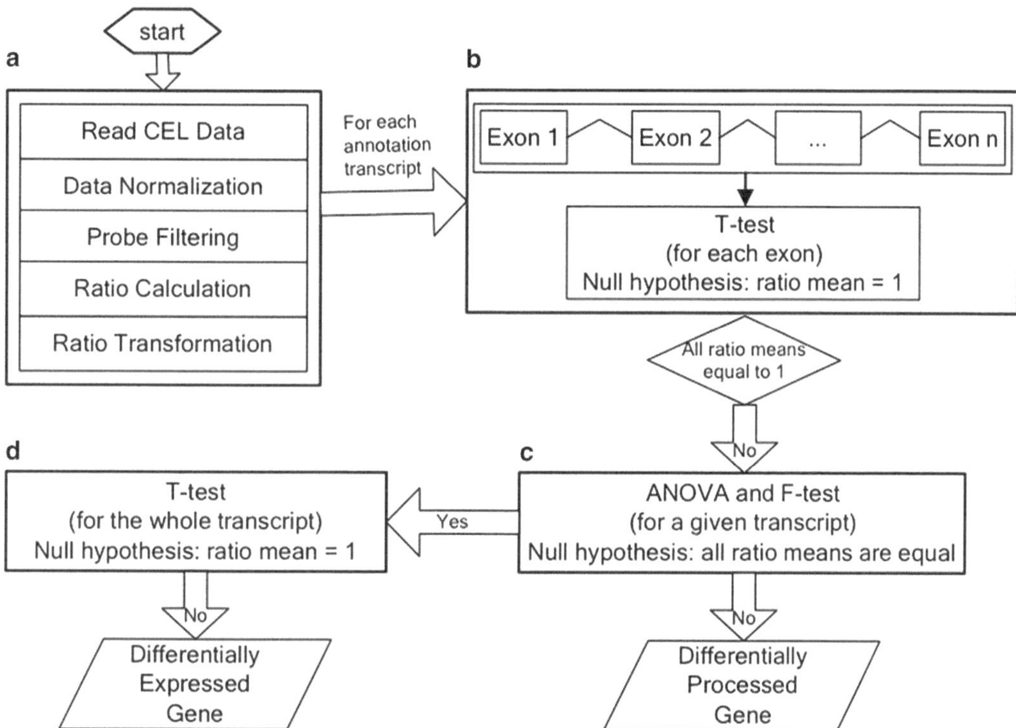

Fig. 2 Schematic representation of RADPRE analysis pipeline. (**a**) Preprocessing of data including background correction, across-array normalization, probe filtering and trimming, ratio generation, and log-transformation. (**b**) To identify transcripts with at least one of its exon ratio means not equal to one, a one-sample two-tails *T*-test was applied to every exon of an annotated transcript with the null hypothesis that the ratio mean of the exon was equal to one. (**c**) For those transcripts identified from the *T*-test in (**b**), a one-way ANOVA and F-test was performed for each transcript with its exons as the "level" parameter. Every transcript with the ratio means of all its exons being equal would be a putative DEG target. Otherwise, the transcript would be a direct DPG target. (**d**) A further one-sample two-tails *T*-test was applied to every one of the putative DEG targets from (**c**) to test whether the ratio mean of the whole transcript was equal to 1. If the ratio mean was not equal to 1, the transcript would be a DEG target. Reused from ref. 5 as permitted by the copyright statement in the original publication

By doing so, one of the major problems intrinsic to tiling array data—the variation of probe affinities—was significantly alleviated. In the second step, RADPRE took the probe ratios as inputs and performed a hierarchy of statistical tests to identify differentially processed genes (DPG) and differentially expressed genes (DEG) (Fig. 2b–d). In the final step, the false discovery rate (FDR) of the DEG and DPG was estimated by using the balanced random combinations of the replicates between two conditions. The output of RADPRE was a few tabular text files with one of them containing the DPG, one the DEG and the others as intermediate files.

While originally designed to identify the APA events in Arabidopsis PCFS4 mutants, RADPRE per se could not differentiate the nature of a given DPG event, be it an APA, an alternative transcription start point or alternative splicing [5]. In another word, the nature of the event has to be determined by the expression profile of the gene involved individually. RADPRE may be adapted for species other than Arabidopsis if the species in question has a completed genome annotation. RADPRE had been successfully applied to identify the DPG targets of Arabidopsis PCFS4 [5]. In this chapter, we use the PCFS4 tiling array datasets as an example to illustrate how to use RADPRE to identify the differential processing events.

2 Materials

2.1 Equipment

1. RADEPRE package (Download source *see* Subheading 2.3).
2. Hardware: 64-bit computer running Linux, Mac OS X (10.6 or later) or MS Windows 7; 2 GB of RAM (4 GB preferred).

2.2 Test Data (Included in the RADPRE Package)

1. CEL files: wt1, wt2, wt3, pcfs4.1, pcfs4.2, and pcfs4.3.
2. CDF files: AtTA.cdf, bpmap.cdf (*see* **Note 1**).

2.3 Setting Up

1. Download RADPRE from the following links: http://www.users.miamioh.edu/liq; or http://www.polyA.org.
2. Install RADPRE on MS Windows:
   Choose "Install Packages" from the Packages menu.
   Select "Install package from local Zip file".
   Select package "Win_RADPRE_2.0.zip".
   Then use the *library*(*RADPRE*) function to load it for use.
3. Install RADPRE on Linux:
   Download "Linux_RADPRE_2.0.tar.gz".
   At the command prompt, install it using:

   ```
   $ R CMD INSTALL/RADPRE_package_path/Linux_
   RADPRE_2.0.tar.gz
   ```

   Use the *library*(*RADPRE*) function within R to load it for use in the session.
4. Install RADPRE on MAC:
   Download "Mac_RADPRE_2.0.tgz". Unzip the package by double-clicking "Mac_RADPRE_2.0.tgz" and put the package under your preferred folder.
5. Open the R console and set your working directory to the folder containing RADPRE package. Load the package "RADPRE" by the following R script to R console:

   ```
   >library(RADPRE)
   ```

3   Methods

In this section, all the R scripts were italic and started with ">". The meaning of each R script was detailed following each segment of the R scripts.

3.1   Read the CEL Files into AffyBatch Objects

```
> celpath = /the file path of the folder contain-
    ing CEL files/
>    cel.filenames=c('R1_080907.CEL','R4_080907.
    CEL','R7_080907_2.CEL','R2_081007.
    CEL','R5_081007.CEL','R8_081007.CEL')
> cel.types=c(rep('wt', 3), rep('pcfs4', 3))
> afbatch.cel=cel.read(cel.filenames, cel.types,
    celfile.path=celpath)
>    sampleNames(afbatch.cel)=c('wt.1',   'wt.2',
    'wt.3', 'pcfs4.1', 'pcfs4.2', 'pcfs4.3')
> pData(afbatch.cel)
> head(exprs(afbatch.cel))
```

celpath: get the file path of CEL filescelpath; *cel.filenames*: select the CEL files you intend to analyze; *cel.types*: *define the file type as control "wt" or treatment "pcfs4"*; *afbatch.cel*: read CEL files into an AffyBatch object (if CEL files path is the current working directory, the parameter "celfile.path" may be omitted); *sampleNames(afbatch.cel)*: Assign the sample name for each of the AffyBatch objects, Optional; pData(*afbatch.cel*): view the affyBatch objects, optional; *head(exprs(afbatch.cel))*: View part of the probe intensity data of each replicate, optional.

3.2   Background Correction and Across-Array Normalization

For background correction and cross-array normalization, "VSN" was performed. You may also choose "RMA" for the processing of the raw data [6, 7].

```
> data(bpmap.cdf)
> afbatch.vsn=data.normalize(RNA.afbatch=afbatch.
    cel, CDF=bpmap.cdf,method='vsn', subsample=0L)
> head(exprs(afbatch.vsn))
```

data(bpmap.cdf): load CDF file with all probes from BPMap file; *afbatch.vsn*: vsn normalization for all probes using subsample data to estimate vsn model parameters (*see* **Note 2**). *head(exprs(afbatch.vsn))*: view part of the normalized and background-corrected probe.

Optional: if you choose to perform the normalization using RMA method, you may run the following code series:

```
> data(bpmap.cdf)

> afbatch.rma=data.normalize(RNA.afbatch=afbatch.
    cel, CDF=bpmap.cdf, method='rma')
```

afbatch.rma: RMA normalization (*see* **Note 3**).

3.3 Ratio Calculation and Transformation

Use the function ratio.calculate.log2 to generate ratios and log-transform the ratios base 2. This function works on probes of each transcript per running.

```
> data(AtTA.cdf)

> type.ctrl='wt'

> isTrt=!(afbatch.vsn$type %in% type.ctrl)

> isCtrl=afbatch.vsn$type %in% type.ctrl

> intensity.trt=exprs(afbatch.vsn[,isTrt])

> intensity.ctrl=exprs(afbatch.vsn[,isCtrl])

> pm_mm_feature=get('AT5G52910.1', envir=AtTA.
    cdf)

> ratio_pm_feature=ratio.calculate.log2(intensity.
    trt, intensity.ctrl,

    gene.cdf=pm_mm_feature,

    is.ratio.paired=TRUE,

    is.input.log2=FALSE)
```

data(*AtTA.cdf*): load CDF files containing probe information and the features of their corresponding annotation unit; *type.ctrl*: designate the control object; *intensity.trt* and *intensity.ctrl*: extract the probe intensities for Treatment and Control; **ratio.calculate.log2**: ratio calculation and log (base 2) transformation of the ratios for each probe (*see* **Note 4**).

3.4 Statistic Tests to Identify DPGs and DEGs

We use the "VSN" normalized data to do the following analysis. The object of "afbatch.vsn" saved the all probe-intensities (*see* Subheading 3.3). The function for Statistic tests (*T*-test, *F*-test) is t.f.tests().

```
> geneID.ls=ls(AtTA.cdf)

> is.ratio.paired=TRUE

> is.input.log2=FALSE

> threshold_ttest.exon=0.01

> threshold_ftest=0.05
```

```
> result=t.f.tests(intensity.trt, intensity.ctrl,
    geneID.list=geneID.ls,
    tilingAnno.cdf=AtTA.cdf,
    is.ratio.paired=is.ratio.paired,
    is.input.log2=is.input.log2,
    threshold.t1=threshold_ttest.exon,
    threshold.f=threshold_ftest,
    threshold.t2=0.05)
> head(result$DPG)
> head(result$DEG)
```

geneID.ls: get the list of genes to be analyzed; *threshold_ttest. exon*: set threshold for *T*-test on each exon; *threshold_ftest*: set threshold for *F*-test on a given transcript (*see* **Note 5**); *result*: The main function to identify the DPGs and DEGs.

3.5 Calculate the False Discovery Rate (FDR)

To estimate the FDR for the identified DPGs ad DEGs, the concept of using balanced random combination of samples from two conditions was applied [8, 9]. Specifically, each balanced random sample constitutes two WT/*pcfs4* and two *pcfs4*/WT ratios. Since we have only three replicates for each of control and treatment, a fourth replicate was simulated by taking the geometric average of three replicates for control and *treatment*, respectively. With the RADPRE analysis of each random combination, a certain number of genes will be called significant. The average number of significant genes from all possible combinations was served as the number of falsely discovered genes. FDR is the ratio of the number of false discovered genes to that of identified DPG and DEG genes.

Note that, for each probe, the intensity of the fourth sample is related to that of the other three samples. So, the relevant degree of freedom (DF) of *T*-test or *F*-test remains to be 3, instead of 4. Two lines of R script within the *t.f.tests*() function, "*is. df.change=TRUE*" and "*df.change=c(n.old=4,n.new=3)*", *serve to adjust the DF*.

```
> Trt.4=2^rowMeans(log2(intensity.trt))
> Ctrl.4=2^rowMeans(log2(intensity.col))
>intensity.T=cbind(intensity.ctrl[,c(1,2)],inten-
    sity.trt[,3],Trt.4)
>intensity.C=cbind(intensity.trt[,c(1,2)],inten-
    sity.ctrl[,3], Ctrl.4)
> result.1=t.f.tests(intensity.T, intensity.C,
```

```
                  geneID.list=geneID.ls,
                  tilingAnno.cdf=AtTA.cdf,
                  is.ratio.paired=TRUE,
                  is.input.log2=FALSE,
                  threshold.t1=threshold_ttest.exon,
                  threshold.f=threshold_ftest,
                  is.df.change=TRUE,
                  df.change=c(n.old=4,n.new=3))
>intensity.T=cbind(intensity.ctrl[,c(1,3)],inten-
    sity.trt[,2],Trt.4)
>intensity.C=cbind(intensity.trt[,c(1,3)],inten-
    sity.col[,2], Ctrl.4)
> result.2=t.f.tests(intensity.T, intensity.C,
                  geneID.list=geneID.ls,
                  tilingAnno.cdf=AtTA.cdf,
                  is.ratio.paired=TRUE,
                  is.input.log2=FALSE,
                  threshold.t1=threshold_ttest.exon,
                  threshold.f=threshold_ftest,
                  is.df.change=TRUE,
                  df.change=c(n.old=4,n.new=3))
> intensity.T=cbind(intensity.ctrl[,1], Ctrl.4,
    intensity.trt[,c(2,3)])
>intensity.C=cbind(intensity.trt[,1],Trt.4,inten-
    sity.col[,c(2,3)])
> result.3=t.f.tests(intensity.T, intensity.C,
                  geneID.list=geneID.ls,
                  tilingAnno.cdf=AtTA.cdf,
                  is.ratio.paired=TRUE,
                  is.input.log2=FALSE,
                  threshold.t1=threshold_ttest.exon,
                  threshold.f=threshold_ftest,
                  is.df.change=TRUE,
                  df.change=c(n.old=4,n.new=3))
> fdr.DPG=mean(nrow(result.1$DPG), nrow(result.
    2$DPG), nrow(result.3$DPG))/nrow(result$DPG)
```

```
> fdr.DEG=mean(nrow(result.1$DEG), nrow(result.
  2$DEG), nrow(result.3$DEG))/nrow(result$DEG)
```

Trt.4: *Simulating the fourth replicate of treatment*; Ctrl.4: *Simulating the fourth replicate of control*; ***intensity.T***: forming the balanced random combination as treatment; ***intensity.C***: forming the balanced random combination as control; result.1, ***result.2***, and ***result.3***: estimation of the false discovered genes for the first, second, and third random combinations, respectively; ***fdr.DPG*** and ***fdr.DEG***: calculating the FDR for DPG and DEG, respectively.

4   Notes

1. When analyzing your own tiling array data, make sure that your CEL files were downloaded to the "CEL files" folder within RADPRE package.

2. The default subsample parameter is set as "*0L*", meaning that all probe-intensities are used for the model parameter estimation. Instead of choosing all probe intensities "0L", you may choose a subset of probes, say "30000". The setting will be "subsample = 30000L". Using a subset of probes will speed up the computation. However, the normalized output data might be slightly changed across multiple runs.

3. If a DNA input is available as a reference for the normalization, the code may be modified as: > *afbatch.rma=data. normalize(RNA.afbatch=afbatch.cel, CDF–bpmap.cdf, method– 'rma', is.DNA.reference=TRUE, DNA.afbatch=DNA.afbatch)*

4. If your experiment was a "Complete Random" design, you may set the "is.ratio.paired" as FALSE. Then, the ratios of randomly paired replicates between Control and Treatment will be calculated.

5. You may change the significance level for either of *T*-test or *F*-test.

Acknowledgement

The authors appreciate the original contributions of Jianti Zheng and Guoli Ji. The project was supported by grants from US National Science Foundation (grant nos. IOS–0817829 and IOS-1353354 to QQL), and from Ohio Plant Biotech Consortium (to QQL and DX).

References

1. Jan CH, Friedman RC, Ruby JG, Bartel DP (2011) Formation, regulation and evolution of Caenorhabditis elegans 3′UTRs. Nature 469(7328):97–101. doi:10.1038/nature09616

2. Wu X, Liu M, Downie B, Liang C, Ji G, Li QQ, Hunt AG (2011) Genome-wide landscape of polyadenylation in Arabidopsis provides evidence for extensive alternative polyadenylation. Proc Natl Acad Sci U S A 108(30):12533–12538. doi:10.1073/pnas.1019732108

3. Marquez Y, Brown JW, Simpson C, Barta A, Kalyna M (2012) Transcriptome survey reveals increased complexity of the alternative splicing landscape in Arabidopsis. Genome Res 22(6): 1184–1195. doi:10.1101/gr.134106.111

4. Thomas PE, Wu X, Liu M, Gaffney B, Ji G, Li QQ, Hunt AG (2012) Genome-wide control of polyadenylation site choice by CPSF30 in Arabidopsis. Plant Cell 24(11):4376–4388. doi:10.1105/tpc.112.096107

5. Zheng J, Xing D, Wu X, Shen Y, Kroll DM, Ji G, Li QQ (2011) Ratio-based analysis of differential mRNA processing and expression of a polyadenylation factor mutant pcfs4 using arabidopsis tiling microarray. PLoS One 6(2):e14719. doi:10.1371/journal.pone.0014719

6. Huber W, von Heydebreck A, Sultmann H, Poustka A, Vingron M (2002) Variance stabilization applied to microarray data calibration and to the quantification of differential expression. Bioinformatics 18(Suppl 1):S96–S104

7. Irizarry RA, Hobbs B, Collin F, Beazer-Barclay YD, Antonellis KJ, Scherf U, Speed TP (2003) Exploration, normalization, and summaries of high density oligonucleotide array probe level data. Biostatistics 4(2):249–264. doi:10.1093/biostatistics/4.2.249

8. Tusher VG, Tibshirani R, Chu G (2001) Significance analysis of microarrays applied to the ionizing radiation response. Proc Natl Acad Sci U S A 98(9):5116–5121. doi:10.1073/pnas.091062498

9. Jones-Rhoades MW, Borevitz JO, Preuss D (2007) Genome-wide expression profiling of the Arabidopsis female gametophyte identifies families of small, secreted proteins. PLoS Genet 3(10):1848–1861. doi:10.1371/journal.pgen.0030171

Part II

Biochemistry of Polyadenylation in Plants

Chapter 7

Characterization of Plant Polyadenylation Complexes by Using Tandem Affinity Purification

Hongwei Zhao, Xinfu Ye, and Qingshun Quinn Li

Abstract

Messenger RNA in eukaryotic cells is initially produced as a nascent transcript (pre-mRNA) without a polyadenine [poly(A)] tail to the 3′ end. The precise cleavage of the pre-mRNA and addition of a poly(A) track need the communication between cis-elements in the pre-mRNA sequences and transacting protein factors recognizing them. Based on homology analyses, Arabidopsis cleavage and polyadenylation specificity factor (AtCPSF) complex should play a critical role in pre-mRNA 3′ end processing. Here we describe the isolation of AtCPSF complex by using a tandem affinity purification (TAP) method. We demonstrate that TAP is a potent protein complex isolating approach that can fulfill a downstream protein identification purpose based on mass spectrometry techniques.

Key words Pre-mRNA, Polyadenylation factor, Tandem affinity purification, Calmodulin binding protein, TEV, Protein purification, Mass spectrometry

1 Introduction

Almost all eukaryotic pre-messenger RNA (pre-mRNA) must undergo a cleavage and polyadenylation procession that produces a mature and functional mRNA with a stretch of poly(A) tail to the 3′ end [1–3]. The site of poly(A) addition on the pre-mRNA is determined by the communication between the cis-elements in the pre-mRNA and the transacting factors that can specifically recognize these cis-elements [1]. The conservation of cis-elements and transacting factors has been frequently utilized in the study of mammalian and yeast polyadenylation machineries. Evidence showed that most of the polyadenylation machinery homologs characterized in other kingdoms can also be found in plants, and it displays an ongoing evolutionary trend from lower to higher species [4–6]. Therefore, it is reasonable to initiate a study by identifying proteins associated with a conserved plant factor and investigates the similarities and differences of pre-mRNA cleavage and polyadenylation

Arthur G. Hunt and Qingshun Quinn Li (eds.), *Polyadenylation in Plants: Methods and Protocols*, Methods in Molecular Biology, vol. 1255, DOI 10.1007/978-1-4939-2175-1_7, © Springer Science+Business Media New York 2015

machinery among the kingdoms. In yeast and animals, groups of proteins such as cleavage and polyadenylation specificity factors (CPSF), cleavage stimulatory factors (CstF), and poly(A) polymerase (PAP) have been identified to participate in pre-mRNA processing [3, 7]. The identity of their counterparts in plants were to be revealed and experimentally verified [8].

Tandem affinity purification (TAP) is a creative protein purification approach that employs two consecutive affinity epitopes. Ideally the two different kinds of affinity tags can be any combination that facilitates efficient protein purification. Sometimes there are enzyme cleavage sites between the two affinity tags so that proteins purified by the first affinity tag can be released (Fig. 1). The first TAP was described by Rigaut and coworkers in 1999 that used a Protein-A and CBP (calmodulin binding protein) combination to successfully purify yeast proteins associated with a bait protein [9, 10]. The Protein-A moiety of the affinity tag originated from a surface protein found in the cell wall of the bacterium *Staphylococcus aureus*. This protein is widely used in biochemical research because of its ability to bind immunoglobulins (IgG). The CBP part can bind calmodulin at a calcium-dependent manner. The TEV protease cleavage site between the two moieties and the reversible nature of the CBP binding ensure that the two consecutive affinity purifications can be released under mild conditions, which allow the use of the second affinity tag. This TAP technique has been successfully utilized in many systems and been modified to either verify single protein-protein interaction or to study high throughput proteomic targets [11]. Here we describe a protocol using TAP to purify Arabidopsis CPSF73-I and its associated proteins for a mass spectrometry (MS) identification of them.

2   Materials

All the solutions used in this protocol are prepared using ultrapure water (prepared by purifying deionized water to attain a sensitivity of 18 MΩ cm at 25 °C). To prevent degradation due to protease activities, all the procedures should be carried out in a cold room or on ice unless indicated otherwise. If the purpose of TAP purification is identifying proteins by using MS, then all the reagents used should be compatible with the downstream MS identification, and should be at or above MS grade. Keep pipette tips and tubes used in this experiment away from direct skin contact, as MS is sensitive enough to pick up trace amount contaminants. Routine chemicals and equipment that are not described in detail in this protocol can be replaced at users' convenience and judgment.

1. Extraction buffer (1×): 20 mM Tris/HCl (pH 8.0), 150 mM NaCl, 0.1 % NP40, 2.5 mM EDTA, 10 mM β-mercaptoethanol, 1 mM phenylmethylsulfonyl fluoride (PMSF), 2 mM benzamide, 20 mM NaF, and 0.1 % (V/V) of protease inhibitors cocktail.

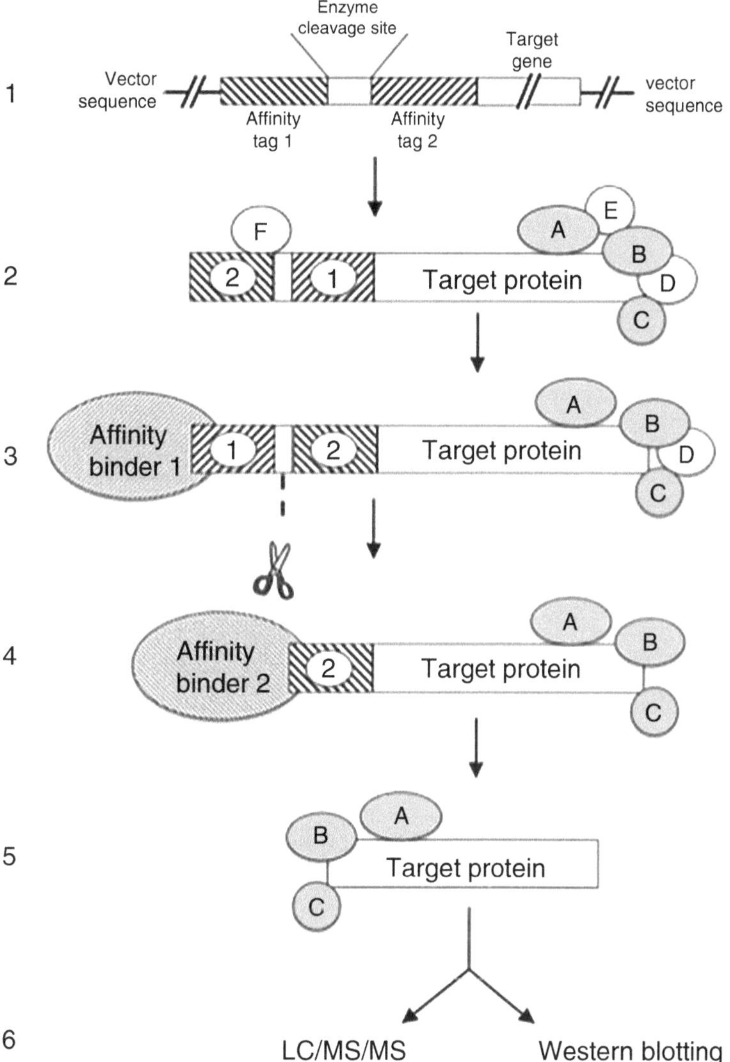

Fig. 1 Schematics of tandem affinity purification (TAP). The methodology of using TAP technique for protein complex purification is outlined in 6 steps. (1) The target protein (the "bait") is co-transcribed with the tandem-arranged affinity tags to either the N- or the C-terminus of the gene (shown is the N-terminal fusion). The two affinity tags (*shaded boxes 1* and *2*) are separated by a protease cleavage site (*blank box*), from where the far-end tag would be cut off and the near-end one would be exposed to the affinity columns for a second-round purification. (2) The "bait" and the tags are expressed as a chimeric fusion protein. The "bait" would form a complex with its in vivo partners (depicted as *A*, *B*, and *C*). Nonspecific association may happen as well, as depicted by *D*, *E*, and *F*. (3) The first round purification via specific interaction between the affinity binder 1 and the far-end tag eliminates most nonspecific associated proteins. (4) In the second round purification, the near-end tag is cut off and the protein complex is further purified, with most nonspecific associates removed. (5 and 6) The "bait" protein and its associated proteins are eluted from the column and are further analyzed either by using antibodies, or by mass spectrometry

2. Immuno-precipitation-150 (IPP-150) buffer: 10 mM Tris/HCl (pH 8.0), 150 mM NaCl, 0.1 % NP40, 1 mM PMSF, and 0.1 % (V/V) protease inhibitor cocktail.

3. Tobacco etch virus (TEV) protease cleavage buffer: 10 mM Tris/HCl (pH 8.0), 150 mM NaCl, 0.1 % NP40, 0.5 mM EDTA, 1 mM DTT, 1 mM PMSF, and 0.1 % (V/V) protease inhibitor cocktail.

4. Calmodulin binding buffer: 10 mM Tris/HCl (pH 8.0), 150 mM NaCl, 1 mM Mg-acetate, 1 mM imidazole, 2 mM $CaCl_2$, 0.1 % NP40 (V/V), 10 mM β-mercaptoethanol, 1 mM PMSF, and 0.1 % (V/V) protease inhibitor cocktail.

5. Calmodulin elution buffer: 10 mM Tris/HCl (pH 8.0), 150 mM NaCl, 1 mM Mg-acetate, 1 mM imidazole, 2 mM ethylene glycol tetraacetic acid (EGTA), 0.1 % NP40 (V/V), 10 mM β-mercaptoethanol, 1 mM PMSF, and 0.1 % (V/V) protease inhibitor cocktail.

6. Chemicals:

   pENTR Directional TOPO Cloning Kit.

   Gateway LR Clonase II Enzyme Mix.

   IgG beads (Immunoglobulin G).

   Tobacco etch virus protease.

   Calmodulin affinity beads.

   Trichloroacetic acid (TCA).

   Sodium deoxycholate (DOC).

   Acetonitrile (ACN).

   NH_4HCO_3.

   Dithiothreitol (DTT).

   Iodoacetamide.

   Sequence grade trypsin.

   Trifluoroacetic acid (TFA).

7. Equipment:

   MicroPulser Electroporation Apparatus.

   SpeedVac.

   Mini columns.

3 Methods

3.1 Construction of Vector for TAP- Fusion Protein Expression in Arabidopsis

1. Follow the instruction provided by the manufacturer (pENTR Directional TOPO Cloning Kit), amplify Arabidopsis CPSF73-I using the following primer set (Forward 5′-CACC ATG GCT TCT TCT TCT ACT TCT CTG AAA AG-3′; Reverse 5′-CTA AGA AGC TGA GAG AGG GAT TGG-3′) [8] (*see* **Note 1**).

2. Set up a PCR system containing:

10 µl	5× PCR reaction buffer
4 µl	dNTPs (2.5 mM)
1 µl	forward primer (10 µM)
1 µl	reverse primer (10 µM)
1 µl	cDNA
0.5 µl	Taq DNA polymerase
32.5 µl	ddH$_2$O

Using a standard thermocycler, amplify the AtCPSF73-I fragment under following conditions:

3 min	94 °C
30 s	94 °C
30 s	55 °C
2 min	72 °C
Repeat **steps 2–4** for additional 29 times	
10 min	72 °C
∞	4 °C

3. Separate the PCR product on a 1 % agarose gel and purify the fragment (around 2,800 bp) by using a gel purification column following the manufacturer's instruction

4. Clone the gel-purified fragment into a pENTR/D vector following the manufacturer's direction.

5. Select the positive colonies by colony PCR.

6. Verify the true transformants by DNA sequencing.

7. Fuse the purified Arabidopsis CPSF73-I fragment with the N-TAPi tag vector (provided Dr. Michael Fromm, University of Nebraska-Lincoln) [12] by using a LR recombination reaction (Gateway LR Clonase II Enzyme Mix) following the manufacturer's suggestion (*see* **Note 2**). A construct expressing the *glucuronidase* protein (GUS) fused with a N-TAPi tag was used as a negative control.

8. Verify the successful fusions by using colony PCRs and extract plasmids from verified colonies for DNA sequencing.

9. Using plasmids containing the genes encoding the TAP-fused proteins, transform *Agrobacterium* strain GV3505 by electroporation by following the manufacturer's instruction.

10. Select the positive transformants by using LB medium containing 40 μg/ml tetracycline, 25 μg/ml rifampin, and 100 μg/ml gentamycin.

11. Transform Arabidopsis plants by using florescence dipping [13]. Select positive plants by using BASTA selection.

3.2 TAP Purification

1. With pre-chilled mortar and pestle (keep chilled to avoid thawing), grind fresh plant tissue or suspension cultured cells in liquid nitrogen to fine powder. For each gram tissue, 1 volume (ml) extraction buffer is added (*see* **Note 3**).

2. Stir the mixture for about 20 min in a cold room.

3. Pass the homogenate through 4-layer cheesecloth pre-wetted with extraction buffer.

4. Spin at $10,000 \times g$ for 10 min at 4 °C.

5. Pass the supernatant through a 60-μm cell strainer (*see* **Note 4**).

6. Transfer the supernatant to a pre-chilled tube containing 100 μl (bed volume) IgG beads. Mix and incubate at 4 °C for 2–4 h (*see* **Note 5**).

7. Collect the IgG beads by passing through a mini column.

8. Wash the beads with cold 10 ml IPP-150 buffer for three times in a cold room.

9. Wash the beads with 10 ml cold TEV protease cleavage buffer for one time.

10. Add 1 ml TEV protease cleavage buffer, 10 μl 0.1 M DTT, and 10 μl (100 units) TEV protease to the mini column. Cap the column tightly and mix by gentle flicks and inversion.

11. Incubate the column (containing beads with the TEV enzyme) in a cold room overnight with constant gentle inversion.

12. Pass the TEV protease digested mixture through the mini column and collect the flow-through into a new mini column.

13. Add 100 μl (bed volume) calmodulin affinity beads to the supernatant (*see* **Note 5**).

14. Add 3 ml of calmodulin binding buffer and 3 μl of 1 M calcium chloride to the flow-through from the previous step (IgG binding). Incubated the column in a cold room for 1 h with gentle rotation (*see* **Note 6**).

15. Drain the column. Wash the beads with 10 ml cold IPP-150 for three times.

16. Elute the proteins by using 1–1.5 ml of elution buffer into 200 μl fractions (*see* **Note 7**). The fractions can be monitored by absorbance at 280 nm or by electrophoresis (*see* **Note 8**).

3.3 Protein Concentration

In order to gain enough amount proteins for trypsin digestion and mass spectrometry detection, precipitate eluted proteins by trichloroacetic acid (TCA) and sodium deoxycholate (DOC) before loading onto a sodium dodecyl sulfate polyacrylamide gel electrophoresis (SDS-PAGE).

1. According to the monitoring results, pool the major fractions containing the eluted proteins. Add 1/100 volume of 2 % (W/V) DOC to each volume of proteins. Mix and keep the tubes on ice for 30 min (*see* **Note 9**).

2. Add 100 µl 100 % (W/V) TCA to a final volume of 6 %. Mix and keep on ice for 1 h.

3. Centrifuge the tube at $2,500 \times g$ for 45 min at 4 °C.

4. Decant the supernatant. Wash the pellets with 100 % cold acetone and spin at $2,500 \times g$ for 45 min at 4 °C.

5. Dry the pellets by using a SpeedVac for 1 min (*see* **Note 10**).

6. Dissolve the pellet in 20 µl SDS-PAGE loading buffer. Separate the samples by a 10 % PAGE gel.

7. Transfer the proteins after SDS-PAGE to a polyvinylidene fluoride (PVDF) membrane and detect the proteins by specific antibodies at 1:2,000 dilutions using Western blot procedures, or by Coomassie Blue staining. Alternatively the proteins can be identified by mass spectrometry as will be described in the following paragraph.

3.4 Trypsin Digestion and Mass Spectrometry Analysis

1. Cut the desired bands from the gels after being separated by SDS-PAGE. Chop the bands into tiny fragments and collect fragments into 1.5 ml Eppendorf tubes.

2. Add 300 µl 50 % (V/V) methanol to each tube and vortex the tubes for 15 min at room temperature. Spin and decant the liquid. Repeat once.

3. Add 300 µl 50 % (V/V) acetonitrile (ACN)/50 mM NH_4HCO_3 (V/V; pH 9.0) to the tubes and vortex for 30 min. Spin and decant the liquid.

4. Add 300 µl 50 % ACN/10 mM NH_4HCO_3 and vortex for 30 min. Spin and decant the liquid.

5. Add 500 µl 100 % ACN and vortex for 10 min. Spin and decant the liquid.

6. Dry the tubes in a SpeedVac for 10 min (*see* **Note 11**).

7. Add 100 µl 10 mM DTT/25 mM NH_4HCO_3 to the dried gel and incubated at 56 °C for 1 h.

8. Remove the DTT/NH_4HCO_3 solution. Add 55 mM iodoacetamide in a volume that is enough to merge the gel (ca 25 µl). Incubate the samples at dark for 45 min at room temperature (*see* **Note 12**).

9. Remove iodoacetamide from the samples. Add 500 µl 25 mM NH$_4$HCO$_3$ to the samples and vortex for 10 min.

10. Remove the liquid. Add 100 µl 100 % ACN to the gels and vortex 5 min. The dehydrated samples are ready for trypsin digestion now.

11. Add 250 µl 10 mM NH$_4$HCO$_3$ (pH 9.0) and 0.5 µg sequence grade trypsin to the gels to start digestion. Keep the samples at 37 °C overnight (*see* **Note 13**). Transfer the solution containing the digested proteins to a new tube.

12. Add 250 µl 0.1 % trifluoric acid (TFA)/water to the tubes containing the gel and shake the tubes for 30 min at room temperature. Transfer the solution containing the digested proteins to a new tube.

13. Add 250 µl 0.1 % TFA/30 % ACN (V/V) to the tubes containing the gel and shake the tubes for 30 min at room temperature. Transfer the solution containing the digested proteins to a new tube.

14. Add 250 µl 0.1 % TFA/60 % ACN to the tubes containing the gel and shake the tubes for 30 min at room temperature. Transfer the solution containing the digested proteins to a new tube.

15. Add 250 µl 0.1 % TFA/90 % ACN (V/V) to the tubes containing the gel and shake the tubes for 30 min at room temperature. Transfer the solution containing the digested proteins to a new tube.

16. Pool the digestion into a new tube and dry to minimal volumes by using a SpeedVac. Reconstitute in 10 µl TFA/water if it was overdried.

17. The reconstituted samples can be sent to a MS facility for analysis.

4 Notes

1. For an N-terminal fusion, design the reverse primer with a stop codon; for a C-terminal fusion, the stop codon should be changed so that the TAP tag at the C-terminus can be translated.

2. The TAP tags are provided by a binary vector containing the TAP tags at either the N-terminus (N-TAPi) or the C-terminus (C-TAPi).

3. For a scale of purification that would be enough for protein identification by mass spectrometry, about 30 g fresh tissue or 50 g suspension-cultured cells (fresh weight) should be used. For suspension-cultured cells, grinding with acid washed sand may be necessary.

4. If a 60-μm cell strainer is not available, the supernatant can be centrifuged one more time. The key point is that the supernatant should be completely free from any debris that could interfere with the following column purification.

5. The beads are normally provided in a storage solution containing about 20 % alcohol and other chemicals such as glycerol. Wash the beads with IPP-150 buffer (or other appropriate solutions) for three times by inversion and brief spinning.

6. The binding of CBP to the calmodulin affinity beads is calcium-dependent. Make sure the binding buffer contains 2 mM Ca^{2+}.

7. Since the binding of CBP to the calmodulin affinity beads is calcium-dependent, make sure the Ca^{2+} in the elution buffer is completely chelated by EGTA with its final concentration equals or above 2 mM.

8. We have used the Protein A280 function of a NanoDrop 1000 for monitoring protein elution peaks which consumes only 2 μl elute each time; but when elute is too diluted for a NanoDrop detection, SDS-PAGE with Coomassie Blue staining or Western blotting detection can be used.

9. DOC is routinely added for precipitation of very small amounts of protein with TCA but the mechanism is not very clear. Some believe DOC binds to proteins and increases the interaction with TCA; others propose that DOC is added in base form (sodium salt) and precipitates after addition of the stronger acid TCA, in which it acts as a coprecipitant and helps the protein to precipitate.

10. The acetone washing removes residual TCA from precipitation, which is necessary for the subsequent procedures. Briefly spin the tube to remove any acetone residue remaining on the wall before vacuum drying has been proven a good practice.

11. This is the step that the experiment can be paused. The dried gel may be stored at –20 °C for months before the next step.

12. Iodoacetamide is an alkylating agent used for peptide mapping purposes. It is commonly used to bind covalently with the thiol group of cysteine so the protein cannot form disulfide bonds. Iodoacetamide is unstable and light sensitive. Therefore, solutions should be prepared immediately before use and perform alkylation in the dark.

13. Modified sequencing grade trypsin must be used in this case because native trypsin is subject to autolysis, generating pseudo-trypsin, which exhibits a broadened specificity. The additional peptide fragments produced could interfere with database search and analysis of the mass of fragments detected by mass spectrometry. Sequencing grade trypsin has been manufactured to provide maximum specificity.

Acknowledgement

We received funding support from the US National Science Foundation (grant nos. IOS–0817829 and IOS-1353354 to QQL). QQL received funding support from the Fujian Hundred Talent Plan.

References

1. Mandel CR, Bai Y, Tong L (2008) Protein factors in pre-mRNA 3′-end processing. Cell Mol Life Sci 65(7–8):1099–1122. doi:10.1007/s00018-007-7474-3

2. Zhao J, Hyman L, Moore C (1999) Formation of mRNA 3′ ends in eukaryotes: mechanism, regulation, and interrelationships with other steps in mRNA synthesis. Microbiol Mol Biol Rev 63(2):405–445

3. Moore MJ, Proudfoot NJ (2009) Pre-mRNA processing reaches back to transcription and ahead to translation. Cell 136(4):688–700. doi:10.1016/j.cell.2009.02.001

4. Xing D, Li QQ (2011) Alternative polyadenylation and gene expression regulation in plants. Wiley Interdiscip Rev RNA 2(3):445–458. doi:10.1002/wrna.59

5. Hunt A, Xing D, Li Q (2012) Plant polyadenylation factors: conservation and variety in the polyadenylation complex in plants. BMC Genomics 13(1):641

6. Hunt AG, Xu R, Addepalli B, Rao S, Forbes KP, Meeks LR, Xing D, Mo M, Zhao H, Bandyopadhyay A, Dampanaboina L, Marion A, Von Lanken C, Li QQ (2008) Arabidopsis mRNA polyadenylation machinery: comprehensive analysis of protein-protein interactions and gene expression profiling. BMC Genomics 9:220. doi:10.1186/1471-2164-9-220

7. Shi Y, Di Giammartino DC, Taylor D, Sarkeshik A, Rice WJ, Yates JR 3rd, Frank J, Manley JL (2009) Molecular architecture of the human pre-mRNA 3′ processing complex. Mol Cell 33(3):365–376. doi:10.1016/j.molcel.2008.12.028

8. Zhao H, Xing D, Li QQ (2009) Unique features of plant cleavage and polyadenylation specificity factor revealed by proteomic studies. Plant Physiol 151(3):1546–1556. doi:10.1104/pp. 109.142729

9. Puig O, Caspary F, Rigaut G, Rutz B, Bouveret E, Bragado-Nilsson E, Wilm M, Seraphin B (2001) The tandem affinity purification (TAP) method: a general procedure of protein complex purification. Methods 24(3):218–229. doi:10.1006/meth.2001.1183

10. Rigaut G, Shevchenko A, Rutz B, Wilm M, Mann M, Seraphin B (1999) A generic protein purification method for protein complex characterization and proteome exploration. Nat Biotechnol 17(10):1030–1032. doi:10.1038/13732

11. Fleischer TC, Weaver CM, McAfee KJ, Jennings JL, Link AJ (2006) Systematic identification and functional screens of uncharacterized proteins associated with eukaryotic ribosomal complexes. Genes Dev 20(10):1294–1307. doi:10.1101/gad.1422006

12. Rohila JS, Chen M, Cerny R, Fromm ME (2004) Improved tandem affinity purification tag and methods for isolation of protein heterocomplexes from plants. Plant J 38(1):172–181. doi:10.1111/j.1365-313X.2004.02031.x

13. Zhang X, Henriques R, Lin SS, Niu QW, Chua NH (2006) Agrobacterium-mediated transformation of Arabidopsis thaliana using the floral dip method. Nat Protoc 1(2):641–646. doi:10.1038/nprot.2006.97

Chapter 8

In Vitro Analysis of Cleavage and Polyadenylation in Arabidopsis

Hongwei Zhao and Qingshun Quinn Li

Abstract

In eukaryotes, pre-messenger RNA (pre-mRNA) cleavage and polyadenylation is one of the necessary processing steps that produce a mature and functional mRNA. Regulation on pre-mRNA cleavage and polyadenylation affects other processes such as mRNA translocation, stability, and translation. The process of pre-mRNA cleavage and polyadenylation, and its relationship with RNA splicing and translation, have been extensively studied due to its importance in vivo. A successful in vitro system has provided enormous amount of information to the study of cleavage and polyadenylation in the mammalian and yeast systems. Here, we describe an in vitro pre-mRNA cleavage system that faithfully cleaves pre-mRNA substrate using Arabidopsis cell/tissue cultures.

Key words Messenger RNA processing, Pre-mRNA 3′-end formation, Cleavage and polyadenylation, In vitro assay, Arabidopsis

1   Introduction

In eukaryotic cells, a nascent messenger RNA (mRNA) is produced as precursor (pre-mRNA) that is uncapped at their 5′ ends, with embedded introns, possesses unprocessed 3′ ends transcribed downstream its polyadenylation site [1]. During maturation, a poly(A) tail is added to the polyadenylation site defined by the communication between the cis-elements in the pre-mRNA sequence and the transacting factors that can specifically recognize these cis-elements [2]. The cleavage and polyadenylation of pre-mRNA is entangled with other processes such as 5′ end processing, splicing, export to the cytoplasm [3], and is important for the efficiency and accuracy of gene expression. When cleavage and polyadenylation is impaired, many other crucial biological activities such as transcription termination by RNA Polymerase II, mRNA stability, mRNA export to

Arthur G. Hunt and Qingshun Quinn Li (eds.), *Polyadenylation in Plants: Methods and Protocols*, Methods in Molecular Biology, vol. 1255, DOI 10.1007/978-1-4939-2175-1_8, © Springer Science+Business Media New York 2015

the cytoplasm, and the efficiency of translation, are all affected [4]. Moreover, small RNA processing [5], noncoding RNA degradation [6] are also impacted in cleavage and polyadenylation mutants.

The study of pre-mRNA has been facilitated by successful in vitro cleavage and polyadenylation systems that faithfully mimic the activities of their in vivo counterparts. In mammals, in vitro systems have been used extensively in studying the functions of different protein factors and testing potential cis-elements [7–9]. In yeast, many polyadenylation factors have been successfully studied by using in vitro systems as well [10, 11]. In contrast, a plant in vitro cleavage and polyadenylation system was not so implemented until recently [12]. Here we describe the establishment of a plant in vitro cleavage and polyadenylation system by using Arabidopsis nuclear protein extract. This system can recognize the cis-elements embedded in the pre-mRNA and direct cleavage on the pre-mRNA as preciously seen in vivo. The cleavage product (hence its RNA-protein complex) of this in vitro cleavage system can be recognized by a yeast poly(A) polymerase (yPAP) [12]. This in vitro assay system is very useful for the study of cis-elements and trans-acting factors that are involved in the cleavage of plant pre-mRNA. In combination with poly(A) polymerase from yeast or other eukaryotic source, this system should provide a potent tool for the study and validation of polyadenylation machinery in plants.

2   Materials

Prepare all solutions using ultrapure water (prepared by purifying deionized water to attain a sensitivity of 18 MΩ cm at 25 °C) and analytical grade reagents. Prepare and store all reagents in a cold room and maintain all the procedures in a cold room or on ice unless indicated otherwise. The steps employing radioisotopes should be performed strictly in an isolated space equipped with necessary protective devices. The space and equipment should be regularly monitored according to a standard radioactive working protocol or individual institute's regulation. Routine chemicals and equipment that are not described in detail in this protocol can be replaced at users' convenience and judgment.

2.1 Nuclear Protein Extraction

1. Extraction buffer (1×): 2.0 M hexylene glycol (2-methyl-2, 4-pentanediol), 20 mM PIPES-KOH (pH 7.0), 10 mM MgCl$_2$, 1 % Triton X-100, and 5 mM β-mercaptoethanol (β-ME). Prepare as a 5× stock solution and store without β-ME at 4 °C; add β-ME immediately before use.

2. Gradient buffer (1×): 0.5 M hexylene glycol, 5 mM PIPES-KOH (pH 7.0), 10 mM MgCl$_2$, 5 mM β-ME, and 1 % Triton X-100. Prepare as a 5× stock solution and store without β-ME at 4 °C; add β-ME immediately before use.

3. Nuclei storage buffer: 50 mM Tris–HCl (pH 7.8), 10 mM β-ME, 20 % glycerol, 5 mM $MgCl_2$, and 0.44 M sucrose.

4. Percoll: Prepare 80 and 30 % percoll suspension by mixing 100 % percoll with 5× gradient buffer. Use ddH_2O to make up to desired volume.

5. Low salt buffer: 25 % glycerol (V/V), 20 mM Tris–HCl (pH 7.9), 1.5 mM $MgCl_2$, 20 mM KCl, 0.2 mM EDTA, 0.5 mM DTT, 0.5 mM PMSF, 0.7 mg/ml pepstatin, 0.4 mg/ml leupeptin.

6. High salt buffer: 25 % glycerol (V/V), 20 mM Tris–HCl (pH 7.9), 1.5 mM $MgCl_2$, 800 mM $(NH_4)_2SO_4$, 0.2 mM EDTA, 0.5 mM DTT, 0.5 mM PMSF, 0.7 mg/ml pepstatin, 0.4 mg/ml leupeptin.

7. Dialysis tubing (10,000 Da cutoff).

2.2 In Vitro Transcription

1. 10× SP6 In Vitro Transcription Buffer.

2. SP6 RNA polymerase.

3. 10 mM ATP, 100 mM CTP, 100 mM GTP, 100 mM UTP, 100 mM DTT, RNase-free water.

4. Gel purification columns.

5. α-^{32}P ATP.

6. RNase-Free DNase I.

2.3 In Vitro Cleavage and Polyadenylation

1. GHP (*Glycerol/HEPES/PVA*-poly(vinyl alcohol)) (*see* **Note 1**): Mix 34.5 ml RNase-free H_2O, 5 ml Glycerol (100 %), 500 μl HEPES (1 M; pH 7.9), 10 ml PVA (10 %).

2. Cleavage buffer: Mix 1 ml GHP solution, 1 ml $MgCl_2$ (25 mM), 1 ml ATP (5 mM), and 83.3 μl creatine phosphate (1.5 M).

3. RNaseOut or other RNase inhibitor.

4. Yeast poly(A) polymerase (yPAP).

2.4 Denaturing Polyacrylamide gel Electrophoresis Examination

1. Components: Urea, Acrylamide, Bis-acrylamide, Ammonium persulfate, TEMED.

2. 2× RNA loading buffer: 95 % Formamide, 18 mM EDTA, 0.025 % SDS, trace amounts of xylene cyanol and bromophenol blue.

3. Gel fixation buffer: 10 % acetic acid; 15 % methanol; in TBE.

4. Sequencing Gel Electrophoresis Apparatus.

5. Gel Dryer.

6. Exposure cassette and PhosphorImager Screen, PhosphorImager scanner. X-ray file and developer can be used as replacement of these items.

3   Methods

3.1   Extract Nuclear Proteins from Arabidopsis Ler Suspension-Cultured Cells or Tissues

1. Collect 5–10 g (completely vacuum drained using a porcelain Buchner funnel with filter paper) suspension-cultured Arabidopsis *Ler* wild type cells or 2–4 g fresh tissue.

2. Grind the cells in liquid nitrogen with acid-washed sands or homogenate the fresh tissue with a Corning blender (*see* **Note 2**)

3. Pass through four-layers cheesecloth pre-wetted with nuclei extraction buffer, squeeze gently without forcing too hard.

4. Prepare one tube (6 ml) 80 % and two tubes (6 ml) 30 % percoll per sample (every 30 ml).

5. Add 6 ml 30 % percoll to a round-bottom centrifuge tube using a Pasteur pipette. Add 6 ml 80 % percoll to the bottom of tubes by passing the pipette through the 30 % percoll layer. Finally add filtered sample on the top of the 30 % percoll without disturbing the surface (*see* **Note 3**).

6. Centrifuge at 2,000×g for 30 min at 4 °C using a swing bucket rotor. Move the tubes to ice without disturbing the dark nuclei band at the interface between the 30 and 80 % percoll phases.

7. Draw off the top layer by a pipette connected to vacuum. Perform with extreme care without disrupting the nuclei layer. Using another pipette, remove the dark band containing nuclei to a new tube. The nuclei are very sticky and tend to travel together. Try to minimize the amount of 80 % percoll being moved together with nuclei (*see* **Note 4**).

8. Add 1× gradient buffer to make a total of 10 ml (*see* **Note 5**).

9. Add 6 ml 30 % percoll to the tube by passing the pipette through the diluted nuclei.

10. Centrifuge 2,000×g for 10 min at 4 °C. Gently decant the supernatant as complete as possible.

11. Judge the packed nuclei volume (PNV). Add 1 volume (relative to the PNV) of low salt buffer to resuspend the nuclei pellet completely (*see* **Note 6**) with a brush.

12. Add about 8 volumes (relative to the PNV) of high salt buffer drop-wisely with constant stirring. Stir additional 10 min.

13. Transfer the mixture to a cold Dounce homogenizer and strike ten times (*see* **Note 7**).

14. Transfer the lysate to a beaker and stir for additional 30 min.

15. Spin the lysed nuclei at 13,200×g for 10 min. Transfer the supernatant to a dialysis tubing (10,000 Da cutoff) without carrying over any solid residue. Dialyze the supernatants against 600 ml extraction buffer for three times (at least 4 h each time) at 4 °C.

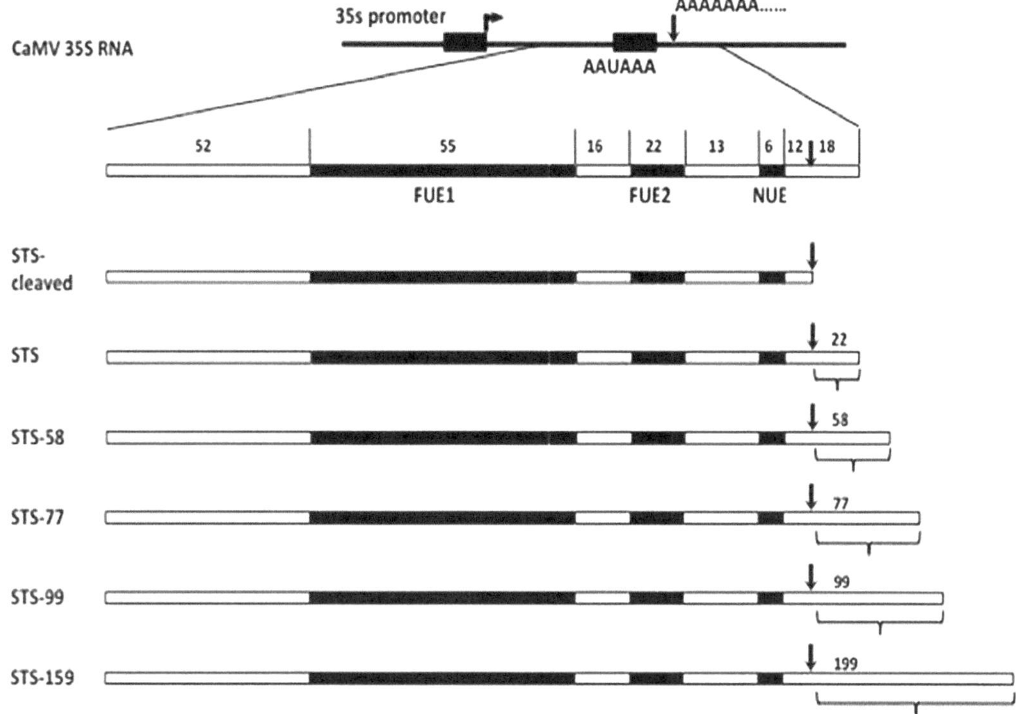

Fig. 1 The RNA used as in vitro cleavage substrates in this study. STS and its extension variants from original sequence are depicted as *bars* with different nucleotide length indicated on the *top*. These fragments can be PCR-amplified from the CaMV 35S 3′-end region. The *black boxes* indicate conserved cis-element (*FUE* far-upstream element, *NUE* near-upstream element, *vertical arrow* cleavage site) while the *white boxes* indicate sequences between them; "AAAAAA   " represents the poly(A) tail

16. Add glycerol to a final concentration of 25 % and keep at −80 °C freezer as aliquots for future uses.

3.2　In Vitro Transcription of Cleavage Substrate

A fragment (termed STS) from the 3′ untranslated region (3′UTR) of the cauliflower mosaic virus (CaMV) 35S RNA is used as the RNA substrate for the plant in vitro cleavage assay (Fig. 1). The STS fragment flanks the polyadenylation site that has been well defined in previous studies [13–15] and has been cloned into a plasmid (pSTS, a gift from Dr. Arthur Hunt, University of Kentucky) for the convenience of this study.

The STS fragment can be amplified by using polymerase chain reactions (PCR) [12]. A 21 base pair (bp) sequence identical to the SP6 promoter (underlined) was designed into the 5′-end of the forward primer (5′-<u>ATTTAGGTGACACTATAGAAC</u>ACGCTG AAATCAC-3′) so that the resulting PCR products would be recognized by the SP6 RNA polymerase and would be transcribed into RNA in vitro. The 3′ primer (5′-GTACTGGATTTTGG TTTTAG-3′) located 20 nucleotides downstream of the cleavage site where the poly(A) tail is added. The substrates produced in this way are labeled by α-^{32}P for the in vitro cleavage assays.

1. Set up a PCR system containing:

2 µl	10× PCR reaction buffer
1 µl	dNTPs
1 µl	SP6 forward primer
1 µl	Reverse primer
0.5 µl	pSTS
0.5 µl	Taq DNA polymerase
14 µl	ddH$_2$O

Using a standard thermo cycler, amplify the STS fragment in following conditions:

3 min	94 °C
0.5 min	94 °C
0.5 min	55 °C
0.5 min	72 °C
Repeat **steps 2–4** for additional 29 times	
10 min	72 °C
∞	4 °C

2. Separate the PCR product on a 3 % agrose gel, excise and purify the STS fragment (around 200 bp) by using a gel purification column following the manufacturer's instruction.

3. With necessary radioactive protection (*see* **Note 8**), assemble a 20-µl in vitro transcription system by combining the following reaction components at room temperature in the order given and mix it well by gently pipetting up and down several times. The components may be kept at 37 °C especially the buffer for optimum results.

1 µl	Purified PCR products with SP6 promoter (100–600 ng)
2 µl	10× AmpliScribe SP6 Reaction Buffer
1.5 µl	10 mM ATP
1.5 µl	100 mM CTP
1.5 µl	100 mM GTP
1.5 µl	100 mM UTP
2 µl	100 mM DTT
6 µl	RNase-Free water
1 µl	α-^{32}P ATP
2 µl	AmpliScribe SP6 Enzyme Solution

Incubate at 37 °C for 2 h.

4. Add 1 μl (1 Unit) of RNase-Free DNase I to the reaction and incubate for 15 min at 37 °C.

5. Heat the reaction for 15 min at 70 °C to inactivate the DNase.

6. Add RNase-free H_2O to a total volume of 200 μl. Add an equal volume of phenol–chloroform–isoamyl alcohol (25:24:1) to the tube, vortex vigorously to mix the phases.

7. Spin in a microcentrifuge at $11,000 \times g$ for 2 min. Remove the aqueous phase to a new tube without carrying over any of the protein between the aqueous and organic phases.

8. Extract the sample with an equal volume of chloroform by vortexing and spinning.

9. Add 1/10 volume sodium acetate (3 M), followed by 2.5 volumes of ethanol and mix well. Incubate at –20 °C for 30 min.

10. Centrifuge at $11,000 \times g$ for 15 min at 4 °C.

11. Remove the supernatant carefully with a pipette and gently rinse the pellet with 70 % ethanol. Air-dry the pellet.

12. Resuspend the RNA pellet in RNase-free water and check the RNA yield by using a spectrophotometer at A_{260}/A_{280} (*see* **Note 9**).

3.3 In Vitro Pre-mRNA Cleavage

1. In a 15 μl system, mix

1 μl	nuclear extract
9 μl	cleavage buffer
Xμl	α-^{32}P labeled CaMV-STS (2,000 CPM, around 2 pmol)
1 μl	RNaseOut (10 Units)
Xμl	RNase-free H_2O (to total of 15 μl)

Gently mix the components by flicking, followed by a brief spin. Incubate the tubes at 30 °C for 3 h.

2. Add 1 μl diluted (1 % dilution, about 6 Units) yPAP (US Biochemical Inc.) to the tube and incubate at 37 °C for 2 h (*see* **Note 10**).

3. Add RNase-free H_2O to a final volume of 100 μl.

4. Add 100 μl of phenol–chloroform and mix by vortexing.

5. Spin at $11,000 \times g$ for 5 min by using a refrigerated centrifuge.

6. Remove 90 μl of aqueous phase to a new tube. Add 9 μl of 3 M sodium acetate and 180 μl cold absolute ethanol. Mix well by inverting six times.

7. Keep at –20 °C for at least 30 min.

8. Spin at $11,000 \times g$ for 15 min by using a refrigerated centrifuge.

9. Wash with 200 µl 70 % ethanol.

10. Air-dry for 10 min.

11. Resuspend the RNA pellet by using 5 µl RNase-free H_2O.

3.4 Denaturing Polyacrylamide Gel Electrophoresis Examination

1. Make a 10 % polyacrylamide sequencing gel containing 7 M urea as described below:

12 ml	ddH_2O
7 ml	5× TBE
14.7 g	urea
3.5 g	acrylamide
98 mg	bis-acrylamide
21 mg	Ammonium persulfate
14 µl	TEMED

Adjust the volume to 35 ml by using ddH_2O. Gently mix the solution by constant stirring and pour the gel immediately (*see* **Note 11**).

2. For each 5 µl RNA samples, add 5 µl of RNA loading buffer, and mixed by gentle flicking.

3. Heat the tubes at 95 °C for 5 min and immediately put the tube on ice to denature the RNA.

4. Load samples to the denaturing urea–polyacrylamide gel.

5. Run the gel at 7 mA constant current till the bromophenol blue runs off the gel (*see* **Note 12**).

6. Fix the gel by immerging into fixation buffer for 15 min.

7. Transfer the gel to 3 M filter paper and dry the gel for 2 h (*see* **Note 13**).

8. In an exposure cassette, develop the dried gel to a PhosphorImager screen for desire exposure.

9. Scan the PhosphorImager screen by using a PhosphorImager scanner (Fig. 2).

10. Analyze the signal strength by using ImageQuant analysis software (**Steps 8–10** may be done by using X-ray film and develop/quantify the film accordingly).

4 Notes

1. The final concentration of each component is: 2 % Glycerol, 6 mM HEPES, 0.4 % PVA, 5 mM Mg^{2+}, 1 mM ATP, 25 mM creatine phosphate, 0.67 Unit/µl RNaseOut, and 10 nM STS RNA with 100 cpm/µl radiation. Dissolving PVA powder to

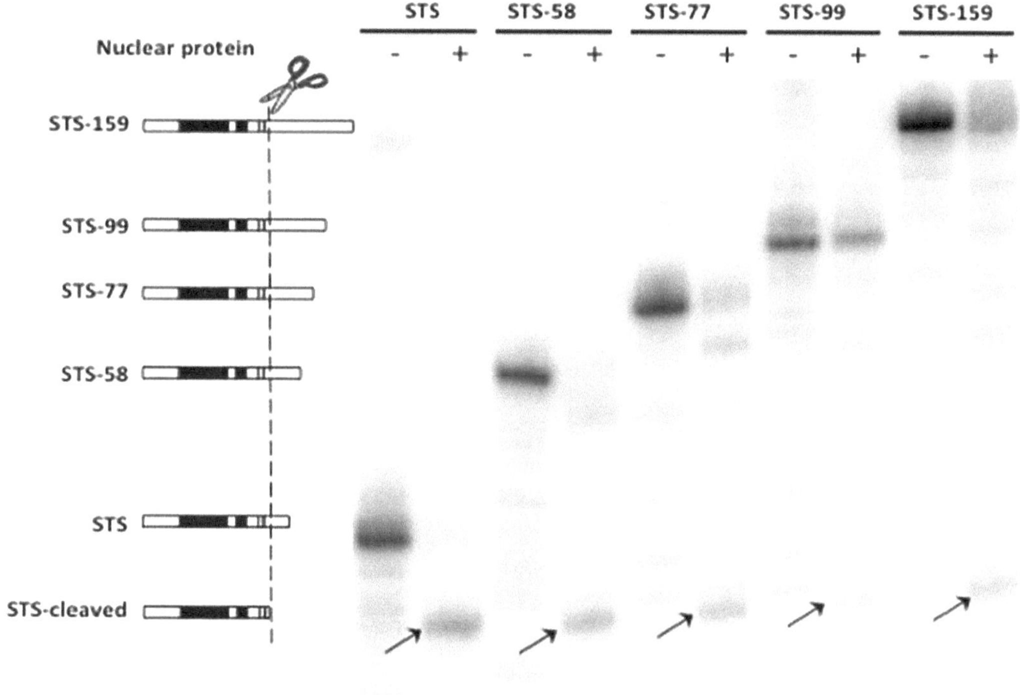

Fig. 2 A typical in vitro cleavage assay is analyzed by PAGE and autophosphorylation. Two picomoles (~2,000 CPM) STS (or its extension variants) were either co-incubated with 1 μl nuclear proteins or extraction buffer under conditions described in Subheading 3. The reactions were separated by a 10 % polyacrylamide sequencing gel (7 M urea). The dried gel was exposed to a PhosphorImager screen and visualized by a Typhoon scanner. The *dashed line* aligned to the *scissors* indicates the cleavage site. *Arrows* point to the correct cleavage products

make a 10 % stock solution needs heating and stirring. Overnight incubation in a 60 °C incubator with constant stirring or inverting proved to be a good procedure.

2. The Corning blender can be replaced by any other blenders with similar functions. The homogenization is quite empirical and needs to be adjusted according to individual equipment used.

3. Attention should be paid not to disturb the border between the 80 and 30 % phases and between samples and the 30 % phase. The latter one can be achieved by controlling the first couple pipetting smoothly and slowly.

4. Use vacuum aspiration if you are a skilled user, otherwise use a Pasteur pipette. Slow down when approaching the interface and remove the 30 % percoll layer as much as possible. Try to limit the 80 % percoll co-transferred as much as possible but small amount is acceptable as it can be eliminated in the following steps.

5. The volume may be up to 20–30 ml if a considerable amount of 80 % percoll was taken in the previous step. The final percoll concentration should be smaller than 30 %.

6. Correctly judge the PNV is needed for adding appropriate volume of lysis buffer in the following steps. If a lysis procedure is not immediately followed, a soft brushed can be used for the resuspension of the intact nuclei pellet in nuclei storage buffer and keep the resuspension at –80 °C for storage.

7. Some protocols using freezing/thaw cycles to break the nuclei. Which method to be used can be decided by pilot experiments. A combination of three freezing/thaw cycles followed by a ten-time Dounce homogenization has been tried and comparable or slightly better results observed.

8. All procedures using or dealing with radioisotope should be conducted in an isolated space and with necessary personal protection unless indicated otherwise.

9. A typical reaction using about 2 pmol of RNA as substrate with 1,000–2,000 cpm/reaction radiation should be sufficient to generate a nice and clear result on a 30×40 cm gel.

10. This step can be skipped if cleavage is the only process examined.

11. Add H_2O and 5× TBE into a 150 ml beaker with stirring (stir speed 3–6 on a stirring plate). Weigh out urea, acrylamide, bis-acrylamide, and AP as listed amount and add to the beaker one by one. No chemicals should be added before the previous one is completely dissolved. Bring to the appropriate volume with ddH_2O. Only minor adjustment needed since the listed volume and amount should give a final volume around 35 ml. Add TEMED, mix gently but thoroughly, and pour gel(s) using a 50 ml syringe immediately. Avoid air bubble during pouring the gels.

12. A piece of metal plate or other cooling system is needed to produce a gel without "smiling effect." Run the gel at low voltage at beginning (200 V for 1 h) followed by 1,000 V would be helpful.

13. If cracking is a problem, try soaking the gel in a solution containing 30 % methanol and 3 % glycerol for 30 min prior drying the gel. For excising the cleavage or polyadenylation products for cloning and sequencing, the gel can be wrapped with a piece of plastic wrap and exposed directly to an X-ray film without drying to identify the location of the bands.

Acknowledgement

We thank Jun Zheng who participated some of the work, and other lab members for suggestions and discussion. This project was supported by U.S. National Science Foundation (grant nos. IOS–0817829 and IOS-1353354 to QQL). QQL received funding support from Xiamen University.

References

1. Zhao J, Hyman L, Moore C (1999) Formation of mRNA 3′ ends in eukaryotes: mechanism, regulation, and interrelationships with other steps in mRNA synthesis. Microbiol Mol Biol Rev 63(2):405–445

2. Xing D, Li QQ (2011) Alternative polyadenylation and gene expression regulation in plants. Wiley Interdiscip Rev RNA 2(3):445–458. doi:10.1002/wrna.59

3. Moore MJ, Proudfoot NJ (2009) Pre-mRNA processing reaches back to transcription and ahead to translation. Cell 136(4):688–700. doi:10.1016/j.cell.2009.02.001

4. Di Giammartino DC, Nishida K, Manley JL (2011) Mechanisms and consequences of alternative polyadenylation. Mol Cell 43(6):853–866. doi:10.1016/j.molcel.2011.08.017

5. Luo Z, Chen Z (2007) Improperly terminated, unpolyadenylated mRNA of sense transgenes is targeted by RDR6-mediated RNA silencing in Arabidopsis. Plant Cell 19(3):943–958. doi:10.1105/tpc.106.045724

6. Lange H, Gagliardi D (2011) Polyadenylation in RNA degradation processes in plants. In: Erdmann VA, Barciszewski J (eds) Non coding RNAs in plants, RNA Technologies. Springer, Berlin, pp 209–225. doi:10.1007/978-3-642-19454-2_13

7. Manley JL (1983) Accurate and specific polyadenylation of mRNA precursors in a soluble whole-cell lysate. Cell 33(2):595–605

8. Moore CL, Sharp PA (1984) Site-specific polyadenylation in a cell-free reaction. Cell 36(3):581–591

9. Ryan K (2007) Pre-mRNA 3′ cleavage is reversibly inhibited in vitro by cleavage factor dephosphorylation. RNA Biol 4(1):26–33

10. Otero LJ, Ashe MP, Sachs AB (1999) The yeast poly(A)-binding protein Pab1p stimulates in vitro poly(A)-dependent and cap-dependent translation by distinct mechanisms. EMBO J 18(11):3153–3163. doi:10.1093/emboj/18.11.3153

11. Dheur S, Nykamp KR, Viphakone N, Swanson MS, Minvielle-Sebastia L (2005) Yeast mRNA Poly(A) tail length control can be reconstituted in vitro in the absence of Pab1p-dependent Poly(A) nuclease activity. J Biol Chem 280(26):24532–24538. doi:10.1074/jbc.M504720200

12. Zhao H, Zheng J, Li QQ (2011) A novel plant in vitro assay system for pre-mRNA cleavage during 3′-end formation. Plant Physiol 157(3):1546–1554. doi:10.1104/pp.111.179465

13. Mogen BD, MacDonald MH, Graybosch R, Hunt AG (1990) Upstream sequences other than AAUAAA are required for efficient messenger RNA 3′-end formation in plants. Plant Cell 2(12):1261–1272

14. Rothnie HM, Reid J, Hohn T (1994) The contribution of AAUAAA and the upstream element UUUGUA to the efficiency of mRNA 3′-end formation in plants. EMBO J 13(9):2200–2210

15. Sanfacon H, Brodmann P, Hohn T (1991) A dissection of the cauliflower mosaic virus polyadenylation signal. Genes Dev 5(1):141–149

Chapter 9

Production, Purification, and Assay of Recombinant Proteins for In Vitro Biochemical Analyses of the Plant Polyadenylation Complex

Stephen A. Bell and Balasubrahmanyam Addepalli

Abstract

In the post-genomic era where gene sequences are available for many organisms, attention has shifted from DNA to the workhorses of the cell—RNA and protein. A number of proteins, as recent studies indicate, seem to possess RNA-binding and RNA cleavage activities. In order to understand the events that comprise RNA processing such as splicing, 3′ end processing, and even RNA turnover, well established methods are necessary. Bacterial recombinant proteins afford an invaluable opportunity to produce proteins in an economical and reproducible fashion in order to study these activities. This chapter describes various experimental protocols to begin the elucidation of the many events that surround RNA processing at the 3′ end of a transcript.

Key words Polyadenylation factor, Affinity purification, Electrophoretic mobility shift assay, Immunoblot analysis, RNA processing, Enzyme assays

1 Introduction

Messenger RNA polyadenylation is an RNA processing event that involves the addition of a polyadenosine [poly(A)] tail to the 3′ end of a pre-mRNA transcript. This process is mediated by a large complex that recognizes specific RNA sequences, processes the pre-mRNA, and adds the distinctive poly(A) tract [1, 2]. It is intertwined with other activities of RNA processing such as the addition of a 5′ methyl guanosine cap and intron/exon splicing [3]. In vitro biochemical analyses have been critical for characterization of the eukaryotic polyadenylation complex and the individual subunits of the machinery [1, 4–6]. Individual subunits have been associated with a plethora of biochemical activities, including RNA and nucleotide binding, protein–protein interactions, and nuclease and nucleotidyltransferase activities, among others. Determinations of

Arthur G. Hunt and Qingshun Quinn Li (eds.), *Polyadenylation in Plants: Methods and Protocols*, Methods in Molecular Biology, vol. 1255, DOI 10.1007/978-1-4939-2175-1_9, © Springer Science+Business Media New York 2015

91

these activities have been very helpful in elucidating aspects of the processing and polyadenylation reaction.

The study of mRNA polyadenylation in plants using biochemical approaches has been particularly challenging, since the complete reaction (cleavage and polyadenylation of a pre-mRNA) is difficult to assay in crude nuclear extracts. To date, processing in crude extracts has been detected, but polyadenylation can only be realized by supplementation with a heterologous (yeast) source of poly(A) polymerase [7]. To overcome this difficulty and dissect the complex interactions and activities of the plant polyadenylation complex, in vitro analyses using homogenous preparations of purified recombinant proteins have been employed. These approaches have provided detailed insight into the various functions of the subunits of the plant complex [8–16].

This chapter will detail protocols used by the authors to study biochemical aspects of the plant polyadenylation complex subunits [8–11]. Three sections are presented for this. The first deals with approaches for overexpression and purification of tagged proteins produced in *E. coli*. The second describes the suite of electrophoretic and associated methods used to assess protein quality and the outcomes of particular biochemical reactions or assays. The third section describes a set of assays that can be performed using the same lysates, purified proteins, and electrophoretic methods to assess protein–protein interactions, RNA binding capability, and nuclease activity.

2   Materials and Equipment

All reagents used should be of the highest grade possible. Impurities found in low-grade reagents used to make buffers and solutions can lead to confounding results. H_2O used to constitute buffers and solutions should at least be distilled and deionized. If possible, water should also be purified with a Nanopure or Milli-Q purification system. In addition, H_2O should either be sterilized by autoclave or passage through a sterile 0.2 μm filter. In lieu of these preparative steps, commercially available, certified RNase-free H_2O may be used.

2.1 Protein Expression, Purification, and Assay Materials

1. Expression vectors encoding fusion proteins bearing coding regions corresponding to the poly(A) factor subunit of interest (*see* **Note 1**).

2. *E. coli* strains BL21(DE3) (New England Biolabs) or Rosetta(DE3) (EMD Millipore) (*see* **Note 2**).

3. Lysogeny Broth (LB—*see* [17]): To 800 mL H_2O, add 10 g Bacto Tryptone, 5 g yeast extract, 10 g NaCl. Stir to dissolve solutes, adjust the pH to 7.0 with NaOH or HCl, and then

bring volume to 1 L. For solid media, add 15 g agar. Sterilize by autoclaving. Cool solid LB to 55 °C before adding antibiotics. Store liquid LB at room temperature, add antibiotics just prior to use. Solid LB plates with antibiotics should be stored at 4 °C prior to use for up to 1 month.

4. Terrific Broth (TB): To 700 mL H_2O, add 12 g Bacto Tryptone, 24 g yeast extract, 10 mL glycerol, and 2 g of glucose (*see* **Note 3**). Stir to dissolve solutes then bring volume to 900 mL. Sterilize by autoclaving. Cool media to room temperature then add 100 mL of filter sterilized 10× phosphate buffer (100 mL: 2.31 g KH_2PO_4, 12.54 g K_2HPO_4). Store liquid TB at room temperature, add antibiotics just prior to use.

5. 1 M Isopropyl β-D-1-thiogalactopyranoside (IPTG).

6. 100 mg/mL ampicillin stock: Dissolve 100 mg of ampicillin powder in 1 mL of sterile H_2O. Store at 4 °C for up to 2 weeks.

7. 100 mg/mL kanamycin stock: Dissolve 100 mg of kanamycin powder in 1 mL of sterile H_2O. Store at 4 °C for up to 2 weeks.

8. 30 mg/mL chloramphenicol stock: Dissolve 30 mg of chloramphenicol powder in 1 mL of ethanol. Store at 4 °C for up to 2 weeks.

9. Protein purification buffer (1×): 50 mM Tris–HCl pH 7.5, 200 mM NaCl, 1 mM ethylenediaminetetraacetic acid (EDTA) (*see* **Note 4**).

10. Column wash buffer (1×): 50 mM Tris–HCl pH 7.5, 2 M NaCl, 1 mM EDTA.

11. Exchange buffer (1×): 50 mM Tris–HCl pH 7.5, 200 mM NaCl.

12. Storage buffer (2×): 100 mM Tris–HCl pH 7.5, 400 mM NaCl, 80 % glycerol.

13. Bio-Rad Poly-Prep columns or Bio-Rad Econo-Columns.

14. Bio-Rad 2-way stopcock and compatible tubing.

15. Amylose resin (New England Biolabs).

16. Glutathione-Sepharose (several sources).

17. Immobilized calmodulin matrix (Agilent).

18. His-Select cobalt affinity gel (Sigma).

19. 1 M NaH_2PO_4 (monobasic, anhydrous): Dissolve 120 g in 1 L of H_2O.

20. 1 M Na_2HPO_4 (dibasic, anhydrous): Dissolve 142 g in 1 L of H_2O.

21. 1 M sodium phosphate buffer, pH 7.5: For 500 mL, mix 80 mL of 1 M NaH_2PO_4 with 420 mL of 1 M Na_2HPO_4.

22. HH purification buffer (1×): 50 mM sodium phosphate buffer pH 7.5, 300 mM NaCl, 5 mM imidazole (*see* **Note 4**).

23. HH wash buffer (1×): 50 mM sodium phosphate buffer pH 7.5, 0.5–2 M NaCl, 5–20 mM imidazole.

24. HH elution buffer: 50 mM sodium phosphate buffer pH 7.5, 300 mM NaCl, 100 mM imidazole.

25. 15 mM MgCl$_2$.

26. Ribonuclease inhibitor (RNasin) (10 U/μL).

2.2 Electrophoretic Materials

1. SDS-PAGE Stacking gel buffer: 0.25 M Tris–HCl pH 6.8.

2. SDS-PAGE Separation gel buffer: 0.75 M Tris–HCl pH 8.8.

3. 30 % acrylamide solution (for SDS-PAGE): 29.2 % acrylamide, 0.8 % N,N'-methylenebisacrylamide.

4. 10 % sodium dodecyl sulfate solution (SDS).

5. 25 % ammonium persulfate solution (APS).

6. N,N,N',N'-Tetramethylethylenediamine (TEMED).

7. Water-saturated butanol.

8. SDS-PAGE loading dye (10×): 100 mM Tris–HCl pH 6.8, 4 % SDS, 1.5 M 2-mercaptoethanol (β-ME), 10 % glycerol.

9. SDS-PAGE running buffer (10×): 250 mM Tris, 1.92 M glycine, 1 % SDS.

10. Coomassie staining solution: 0.5 % Coomassie Brilliant Blue (CBB) R-250, 50 % methanol, 5 % glacial acetic acid.

11. Coomassie destaining solution: 40 % methanol, 10 % glacial acetic acid.

12. Tris-borate-EDTA (TBE) buffer (5×): 445 mM Tris, 445 mM borate, 10 mM EDTA, prepared with RNase-free H$_2$O.

13. Acrylamide (crystalline, molecular biology grade).

14. N,N'-Methylenebisacrylamide (crystalline, molecular biology grade).

15. Ammonium persulfate, solid (APS).

16. EMSA loading dye: 20–30 % glycerol, 0.1 % bromophenol blue, and 0.1 % xylene cyanol in H$_2$O.

17. Urea (solid).

18. Sequencing gel loading dye: 98 % formamide, 10 mM EDTA, 0.1 % bromophenol blue, 0.1 % xylene cyanol in H$_2$O.

19. Transfer tank buffer (1×): 192 mM glycine, 25 mM Tris, 20 % methanol.

20. TTBS (5×): 200 mM Tris–HCl pH 7.5, 2.5 M NaCl, 0.25 % Tween-80.

21. Blocking solution: 25 g nonfat dry milk dissolved in 400 mL of 1× TTBS.

22. Antibodies: Anti-MBP Monoclonal Antibody (New England Biolabs), Anti-Mouse IgG (whole molecule)-Alkaline Phosphatase antibody produced in goat (Sigma-Aldrich), and Monoclonal Anti-polyHistidine–Alkaline Phosphatase antibody produced in mouse (Sigma-Aldrich).

23. 20 mg/mL 5-bromo-4-chloro-3-indolyl phosphate (BCIP) stock: Dissolve 20 mg of BCIP (disodium salt) powder in 1 mL of H_2O.

24. 20 mg/mL Nitro Blue Tetrazolium chloride (NBT): Dissolve 20 mg of NBT powder in 1 mL of 70 % (v/v) dimethylformamide (DMF).

25. Alkaline Phosphatase (AP) buffer (1×): 100 mM Tris–HCl pH 9.5, 100 mM NaCl, 5 mM $MgCl_2$.

26. 0.2 μm Nitrocellulose membrane.

27. 3 mm Whatman paper.

2.3 Equipment

1. Centrifuges (benchtop and floor models, cooled if possible).

2. Water bath or heating blocks.

3. Vertical and horizontal electrophoresis equipment.

4. Gel dryer.

5. X-ray film or phosphor imaging plates.

6. Phosphor imager.

7. Plexiglass screens for protection from radiation.

8. Liquid scintillation counter.

9. Spectrophotometer.

10. Cuvettes and scintillation vials.

11. Amicon Ultra Centrifugal Filter Units.

12. Scintillation Cocktail.

3 Methods

3.1 Protein Expression and Purification

Recombinant, tagged polyadenylation factor subunits are produced in *E. coli* using any of a number of expression vectors (*see* **Note 1**) and specialized *E. coli* strains (*see* **Note 2**). The process for this involves assembly of the recombinant plasmid, screening recombinants for protein production, and production of cleared lysates that may be used for further studies. In the following, a procedure for generating and screening independent clones encoding recombinant proteins is described. Subsequently, a generalized method for purification of tagged proteins is described. For this, it is assumed that the appropriate expression plasmid and cloning strategy have been identified and successfully carried out.

3.1.1 Growth of E. coli Expressing Recombinant Proteins

1. Transform 0.5–1 μL (0.01–0.5 ng) of sequence verified plasmid DNA (or the empty cloning vector, as required by the experiment) into freshly prepared *E. coli* competent cells (*see* **Note 5**) and plate on the appropriate selection media. Incubate at 37 °C overnight (12–15 h) then store the plate at 4 °C for up to 1 week until needed for the next step.

2. From the empty vector control and experimental construct transformation plates, select three biological replicates for protein expression (*see* **Note 6**). Use a 10 μL pipette with a sterile tip to poke a single colony then eject the tip into a culture tube that contains 10 mL of LB selection media (*see* **Note 7**). Grow the cultures overnight at 37 °C with 220 rpm orbital shaking. If possible, slant the culture tube for better aeration of the culture.

3. Use 1 mL of the overnight culture to inoculate 100 mL of Terrific Broth (TB) containing the appropriate antibiotic (for selection of recombinant cells) in a 500 mL flask (*see* **Notes 7 and 8**). Incubate the culture at 37 °C with 220 rpm orbital shaking for 2.5 h.

4. Remove the cultures and cool to room temperature by sitting on ice for 2–3 min with periodic swirling. Induce protein expression by adding 50 μL of 1 M IPTG (0.5 mM final concentration). Grow cultures for 5 h at 23–25 °C with 220 rpm orbital shaking (*see* **Note 9**).

5. Dispense the cultures into 50 mL centrifuge tubes (for large-scale protein purification) or 14 mL centrifuge tubes (10 mL of culture per tube; for small scale production of cleared lysates) and centrifuge at $2,000 \times g$ for 10–20 min. Decant and discard the supernatant then place the tube upside down on a paper towel to remove residual media. Freeze the cell pellets at –80 °C (*see* **Note 10**).

6. Resuspend one of the 10 mL pellets for each biological replicate in 1 mL of the appropriate protein purification buffer (Subheading 2.1). Transfer 100 μL of the suspension to a 0.5 mL tube with 10 μL of 10× SDS-PAGE loading dye. Heat the sample to 98–100 °C for 20 min.

7. Pellet the cellular debris by centrifuging at $10,000 \times g$ for 10 min. Analyze 15–20 μL by SDS-PAGE and immunoblot analyses (described in Subheading 3.2; *see* **Note 11**).

3.1.2 Preparation of Crude Lysates

1. Thaw the cell pellets from **step 5** of Subheading 3.1.1 on ice for ~30 min. Add 5 mL of cold protein purification buffer (*see* **Notes 12 and 13**) for every 50 mL of initial culture volume (Subheading 3.1.1, **step 5**) and resuspend the cells by gently vortexing. Transfer the suspension to a tube that is safe for use with a probe-style sonicator.

2. Lyse the resuspended cells using a probe sonicator (*see* **Note 14**). Insert the probe in the sample ~1.5 cm below the surface and sonicate for 15–20 s, using a power setting that produces a distinctive crackling sound in the suspension. Place the sample on ice for 5 min. Repeat for a total of three sonication bursts.

3. Pellet the insoluble cellular debris by centrifuging the sample at $5,000$–$10,000 \times g$ for 20 min at 4–8 °C (*see* **Note 15**). Remove and save the supernatant; this is the crude cell lysate.

3.1.3 Purification of Affinity-Tagged Proteins

The following is a generic protocol that can be used with a range of affinity-tagged proteins. Importantly, it does not utilize immuno-purification, and thus is not intended for the purification of tagged proteins using antibodies raised against the respective tags. The procedure is amenable for use with disposable column systems as well as with more durable and permanent glass columns. All steps for protein purification should be performed at 4–8 °C. All buffers should be chilled to 4 °C prior to use.

1. Using a ring stand and clamp holder, secure a column in the upright position with a 2-way stopcock and flexible tubing fashioned to the bottom end. Carefully add 2 mL of the appropriate purification matrix (*see* **Note 16**) to a column using a 1,000 µL pipette tip with the end cut off. Wash the column with 10 mL of cold H_2O to remove the storage solution, taking care to avoid drying of the resin. Equilibrate the column by washing with 10 mL of cold purification buffer (*see* **Note 17**).

2. Apply the clarified crude lysate (Subheading 3.1.2, **step 3**) onto the column using a 10 mL serological pipette (*see* **Note 18**). Collect the flow-through in a new tube that is kept on ice. Once the lysate has been loaded onto the top of the resin bed, adjust the flow rate to ~2 mL/min (*see* **Note 19**). When the majority of the lysate has passed through the column, close the stopcock valve. Pass the lysate over the column two more times in the same manner just described.

3. Wash the column with 10 mL of wash buffer using a flow rate of 1–2 mL/min. Then, wash the column with 10 mL of purification buffer (*see* **Note 20.**). After the second wash, leave ~500 µL of buffer above the surface of the bedded resin/gel, if the elution is delayed.

4. Elute the protein, after draining the previous liquid, with 10 mL of the appropriate elution buffer (*see* **Notes 21** and **22**). Collect the elution in a new 14 mL tube that is kept on ice.

5. In an appropriately outfitted centrifuge with a fixed angle rotor that has been chilled to 4–8 °C, load the eluted protein sample onto an Amicon Ultra-15 filter unit and centrifuge at $4,500 \times g$ until the volume reaches ~250 µL (*see* **Note 23**). Add 10 mL of exchange buffer and repeat the centrifugation step again

until the volume reaches ~250 µL. Repeat this step once more to sufficiently dilute the elution buffer.

6. Add an equal volume of the 2× storage buffer to the concentrated sample and store at –20 °C or –80 °C (*see* **Notes 24** and **25.**).

3.2 Electrophoretic Methods

The initial characterizations of the various fusion proteins (Subheading 3.1.1, **step 7**), as well as more detailed biochemical assays (Subheading 3.3) make extensive use of electrophoretic separations and immunoblotting. Protocols used in the authors' laboratories for these various techniques (SDS-polyacrylamide gel electrophoresis, native acrylamide gel electrophoresis, and separation of nucleic acid on acrylamide gels containing urea) are described in this section.

Proper precautions should be taken when working with acrylamide such as gloves, goggles, and lab coats because of its neurotoxic effects. Polymerized acrylamide, while potentially less toxic, should still be handled carefully and disposed of properly.

3.2.1 SDS-Polyacrylamide Gel Electrophoresis (SDS-PAGE)

1. In a 50 mL flask, mix 2.3 mL H_2O, 4.5 mL 30 % acrylamide solution, 7.5 mL SDS-PAGE Separation gel buffer, 150 µL 10 % SDS, 12 µL TEMED, and 50 µL of 25 % APS (Subheading 2.2, items 1-6) (*see* **Note 26**). Thoroughly mix the components by swirling the flask then transfer the liquid into the sealed gel sandwich (*see* **Note 27**) with a 1 mL pipette. Be sure to leave adequate space at the top for the stacking gel and comb. Add a small layer of water-saturated butanol to the top to smooth the interface.

2. After the gel has polymerized, thoroughly flush the water-saturated butanol from the top of the gel by flooding with ddH_2O. Stand the gel on its side so that excess water will pool at the edge of the plates. Use a low-lint wipe to dab the excess water from the gel.

3. In a 50 mL flask, mix 2.9 mL H_2O, 0.75 mL 30 % acrylamide solution, 3.75 mL SDS-PAGE Stacking gel buffer, 75 µL 10 % SDS solution, and 6 µL TEMED. Degas the stacking gel solution by either pulling a vacuum on the flask (*see* **Note 28**) or placing the flask in a water bath sonicator.

4. Add 25 µL of 25 % APS solution and mix the components thoroughly by swirling the flask. Use a 1 mL pipette to transfer the stacking gel solution on top of the separation gel. Carefully insert a comb into the stacking gel making sure that no bubbles form. Allow polymerization to occur at room temperature (*see* **Note 29**).

5. Clamp the gel plates into a gel running apparatus and fill the tank compartment with 1× running buffer. Allow the gel to soak in the buffer for at least 10 min to make sure the buffer does not leak from the tank compartment.

6. Carefully remove the comb from the stacking gel and allow the wells to fill with running buffer. Load the samples using a pipettor and suitable pipette tip (*see* **Note 30**). Run the gel at 100–150 V for approximately 1–1.5 h or until the tracking dye reaches the bottom of the gel.

7. Turn the power supply off, remove the gel plates from the gel running apparatus, and rinse with ddH$_2$O. Carefully split the glass plates apart with a wedge or scalpel blade. Use a scalpel blade to cut along the interface of the stacking and separation gels, and then remove and dispose of the stacking gel. Reserve the gel for further processing (see following sections).

3.2.2 Staining and Destaining SDS-PAGE Gels

1. After the stacking gel is removed (Subheading 3.2.1, **step 7**), carefully peel the gel from the plate and fully submerge it in Coomassie stain in a plastic tray. Place the tray in a microwave (*see* **Note 31**) and microwave on high for 10 s. Open the microwave and gently rock the tray. Repeat this two more times.

2. Allow the gel to incubate at room temperature in a fume hood for ~10 min then pour the Coomassie stain back into a bottle (*see* **Note 32**). Rinse the gel and tray with ddH$_2$O to remove excess dye, then add destaining solution and a piece of foam sponge or low-lint wipe (*see* **Note 33**).

3. Rock gently for 30 min then change destaining solution. Rock for another 30 min. Change destaining solution once more, place plastic wrap and a rubber band around the tray, and gently rock overnight.

4. The next day, remove the destaining solution and soak the gel in ddH$_2$O for several hours. Capture an image of the gel using a scanner or camera.

3.2.3 Immunoblot Analysis

1. After the SDS-PAGE gel has run and the stacking gel has been removed (Subheading 3.2.1, **steps 6** and 7), set up a transfer of the proteins from the SDS-PAGE gel to the nitrocellulose membrane (*see* **Note 34**). Prepare the transfer sandwich, making sure that the sandwich components are placed in the correct order as follows: bottom of sandwich holder–sponge, two pieces of 3 mm Whatman paper, gel, nitrocellulose, two pieces of 3 mm Whatman paper, sponge–top of sandwich holder (*see* **Note 35**).

2. Once the sandwich is constructed, place it in the transfer apparatus in a buffer tank that contains 1× transfer tank buffer. Make sure the bottom side of the transfer sandwich is toward the negative (black) terminal. Transfer the proteins from gel to membrane using current and voltage settings appropriate for the electrotransfer unit.

3. After the transfer is complete, disconnect the power supply (*see* **Note 36**) and remove the transfer sandwich. Carefully break

the sandwich apart and peel the nitrocellulose membrane away from the gel making sure to check that the protein-sizing standard has been transferred to the membrane.

4. Add 1× TTBS to a plastic tray with a smooth bottom and place the membrane in it for 5 min with gentle rocking (*see* **Note 37**). Pour off and discard the TTBS, add Blocking Solution, and rock for 1 h.

5. Pour off the Blocking Solution and add 10 mL of fresh Blocking Solution along with a suitable quantity of primary antibody (*see* **Notes 38** and **39**). Cover the tray with plastic wrap and a rubber band and gently rock overnight at room temperature.

6. Pour off the Blocking Solution and wash three times with 1× TTBS with 5 min of rocking (*see* **Note 37**). Add 10 mL of Blocking Solution and the recommended amount of secondary antibody (*see* **Notes 40** and **41**). Rock for 2 h.

7. Pour off the Blocking Solution and wash three times with 1× TTBS with 5 min of rocking in between (*see* **Note 42**).

8. In a separate plastic tray, add 20 mL of 1× AP buffer, 50 µL of 20 mg/mL NBT solution, and 50 µL of 20 mg/mL BCIP solution (*see* **Note 43**). Mix the developing solution thoroughly by rocking the tray.

9. Quickly rinse the membrane with ddH$_2$O and then place it in the tray with developing solution. Allow the blot to develop until bands appear to a satisfactory intensity with minimal background levels. When the development is complete, flood the tray with ddH$_2$O several times and place the membrane on a paper towel to dry.

3.2.4 Native Polyacrylamide Gel Electrophoresis

1. In a clean 125 mL flask, add 3 g acrylamide, 60 mg *N,N'*-methylenebisacrylamide, 120 mg APS, and 75 mL TBE (*see* **Note 44**). Mix the components thoroughly by swirling the flask.

2. Once the solutes are dissolved, degas the solution (*see* **Note 28**), add 150 µL TEMED and swirl to mix. Rapidly and carefully pour the solution into a sealed gel sandwich (*see* **Note 27**) and insert a comb to form the wells. Allow polymerization to occur at room temperature until complete (*see* **Note 45**).

3. Rinse the gel sandwich with sterile water (to remove trace amounts of unpolymerized acrylamide) and carefully remove the comb. Assemble the gel sandwich into an electrophoresis tank and fill the chambers with 1× TBE. Pre-run the gel for 15 min at room temperature before loading the samples.

4. Load each sample and conduct the electrophoresis at constant voltage until the tracking dye is near the bottom of the gel (*see* **Note 46**).

5. When the electrophoresis has completed, carefully pry the glass plates apart such that the gel remains on one of the glass plates. Lay a dry piece of 3 mm Whatman paper on top of the gel and use a glass rod or pipette to gently roll over the plate so that the gel uniformly adheres to the Whatman paper.

6. Gently peel the Whatman paper and gel from the glass plate being careful not to tear the gel while also making sure the gel is stuck to the Whatman paper. Dry the gel under vacuum at 50–60 °C for 1–2 h.

7. After the gel is dry on the Whatman paper, wrap it completely in plastic wrap. Expose the wrapped, dried gel to an X-ray film or phosphor imaging plate. Develop using an appropriate system.

3.2.5 Sequencing Gel Electrophoresis

1. Mix the following in a 125 mL flask: 14.7 g of urea (7 M final concentration), 2.74 g acrylamide (for 7.5 % polyacrylamide gels) or 5.25 g acrylamide (for 15 % gels), 78 mg or 157 mg bisacrylamide (for 7.5 % or 15 % gels, respectively), 21 mg ammonium persulfate, 7 mL 5× TBE, and 5 mL H_2O (*see* **Note 47**). Heat to 40 °C using a stirring hot plate to dissolve the solutes. Once dissolved, bring the volume to 35 mL with RNase-free H_2O.

2. Cool the mixture to room temperature (*see* **Note 48**) and then add 11 μL of TEMED. Swirl to mix then use a 1 mL pipette to transfer the liquid to the glass plates set up in a casting apparatus (*see* **Note 49**). Add the appropriate comb and allow gels to polymerize at room temperature.

3. After loading the samples run the gel at 8–10 mA constant current until the tracking dye has traveled 90 % of the distance of the gel. Following electrophoresis, transfer the gel to 3 mm Whatman paper and dry under vacuum at 50–60 °C for 1–2 h.

4. After the gel is dry on the Whatman paper, wrap it completely in plastic wrap. Expose the wrapped, dried gel to an X-ray film or phosphor imaging plate. Develop using an appropriate system.

3.3 Biochemical Assay of Plant Polyadenylation Factors

Plant polyadenylation factors possess distinctive and characteristic biochemical activities. Nucleotidyltransferase assays are described in another chapter in this volume. Assays of other common and important activities—protein–protein interactions, RNA binding, and nuclease—are described in this section.

3.3.1 Co-purification Assays to Measure Protein–Protein Interactions

This assay has been used to assess direct protein–protein interactions between different tagged subunits of the polyadenylation complex. It utilizes crude extracts that contain the tagged proteins of interest; each protein being tested will have a different tag to permit differentiation during purification and immunoblot analysis.

For example, MBP and GST fusion proteins may be tested by mixing the respective crude lysates, purifying the MBP fusion protein using amylose resin, and assessing co-purification by immunoblot analysis using anti-GST primary antibodies.

1. Prepare crude lysates containing the appropriate fusion proteins as described in Subheading 3.1.2. Lysates containing control proteins should also be prepared as needed (*see* **Note 50**).

2. Mix 200 µL of the appropriate crude lysate in a 1.5 mL tube. Add any other required substrates or compounds (*see* **Note 51**). Incubate the reactions in a heat block at 30 °C for 5–60 min (*see* **Note 52**).

3. While the lysates are incubating, transfer 100 µL of the appropriate affinity matrix (*see* **Note 16**) to a 1.5 mL tube using a cut pipette tip. Add 1 mL of the appropriate protein purification buffer (*see* **Note 51**), invert several times, and pellet the resin by gentle centrifugation ($500 \times g$, 10 s). Remove the supernatant with a pipette and repeat this wash step. Leave at room temperature uncapped until the next step.

4. After the binding reactions have incubated for the desired times, transfer each 400 µL reaction to the tube with washed affinity resin. Gently flick the tube to mix the resin with the reaction and incubate at room temperature for 10 min (gently flick the tube several times during the incubation).

5. Pellet the resin by gentle centrifugation ($500 \times g$, 10 s). Remove the supernatant with a pipette and add 1 mL of the appropriate protein purification buffer. Pellet the resin, remove and discard the supernatant, and repeat this wash step two more times.

6. Add 100 µL of the appropriate elution buffer (*see* **Note 21**) and incubate at room temperature for 5 min. Pellet the resin by centrifugation and transfer the supernatant to a new tube.

7. Prepare aliquots of the crude lysate (Subheading 3.1.2, **step 3**) and elution (Subheading 3.3.1, **step 6**) for SDS-PAGE by mixing 25 µL of sample with 2.5 µL of 10× SDS-PAGE loading dye (Subheading 2.2, **item 8**). In parallel, add 25 µL of purification buffer and 2.5 µL of SDS-PAGE loading dye to the pelleted resin from Subheading 3.3.1, **step 6**. Boil the samples for 5 min. For the sample with the resin, pellet the beads by gentle centrifugation ($500 \times g$, 10 s).

8. Load and run on SDS-containing polyacrylamide gels (Subheading 3.2.1) and analyze by immunoblotting (Subheading 3.2.3). (*See* **Note 53**.)

3.3.2 RNA Binding Assays

The following protocol describes an assay for detecting RNA–protein interactions involving purified subunits of the plant polyadenylation complex. Purified protein (Subheading 3.1.3, **step 6**) is used in a binding reaction with a suitable radiolabeled RNA. Subsequently, complexes are separated from unbound RNA on native acrylamide gels and detected by autoradiography.

All reagents, solutions, and materials used for RNA transcription and binding should be carefully prepared to prevent contamination by RNases. These precautions include using certified RNase-free H_2O, filter tips, and tubes. Experiments should be conducted in clean working spaces; gloves and lab coats should always be worn. All personnel should be properly certified to work with radioactive isotopes as dictated by the guidelines for the respective research institute where the experiment is conducted. Radioactive work should always be conducted in a clearly labeled and designated area with the proper protective plexiglass barriers in place.

1. Mix 2–20 pmol radiolabeled RNA (*see* **Note 54**), unlabeled RNA (if required; *see* **Note 55**), 2–40 pmol purified protein (Subheading 3.1.3, **step 6**; *see* **Note 56**), 0.4 μL 15 mM $MgCl_2$, and other compounds appropriate for the experiment (*see* **Notes 51** and **57**). Bring the final volume to 10 μL with the protein purification buffer appropriate for the purified protein (e.g., reagent 2.1, **item 9** for CDB, MBP, and GST fusion proteins, reagent 2.1, **item 22** for HH fusion proteins).

2. Incubate the assembled reaction at 25–30 °C for 15 min (*see* **Note 58**). Stop the reaction by adding an equal volume of EMSA loading dye.

3. Using a standard (1–20 μL) pipettor and pipette tip, load the entire sample onto a lane of a native polyacrylamide gel (Subheading 3.2.4). Run the gel and process as described in Subheading 3.2.4. (*See* **Note 59**.)

3.3.3 Ribonuclease, RNA Processing, and Nucleotidyltransferase Assays

This assay is a generic one for measuring modifications of a substrate RNA. It makes use of denaturing gel electrophoresis to provide better resolving power for products of RNA-modifying reactions, thereby allowing for measurement of processing reactions, nucleotidyltransferase activities (as associated with poly(A) polymerases), and exonucleolytic as well as endonucleolytic activities. As with the EMSA method, all reagents, solutions, and materials used for RNA transcription and binding should be carefully prepared to prevent contamination by RNases. These precautions include using certified RNase-free H_2O, filter tips, and tubes. Experiments should be conducted in clean working spaces; gloves and lab coats should always be worn. All personnel should be properly certified to work with radioactive isotopes as dictated by the guidelines for the respective research institute where the experiment is conducted. Radioactive work should always be conducted in a clearly labeled and designated area with the proper protective plexiglass barriers in place.

1. Mix 2–20 pmol radiolabeled RNA (*see* **Note 54**), 2–40 pmol purified protein (*see* **Note 56**), 0.4 µL of 15 mM MgCl$_2$ (0.6 mM final concentration), and other components that need to be tested (*see* **Notes 51** and **60**). Bring the final volume to 10 µL with the protein purification buffer appropriate for the purified protein (e.g., reagent 2.1, **item 9** for CDB, MBP, and GST fusion proteins, reagent 2.1, **item 22** for HH fusion proteins).

2. Incubate the reactions at the appropriate temperature (25–30 °C) for an appropriate time (*see* **Note 61**). Stop the reaction by the addition of an equal volume of phenol–chloroform (1:1) followed by vortexing and centrifugation at 10,000 × *g* for 10 min.

3. Mix 3–5 µL of the aqueous phase with an equal volume of denaturing gel loading dye (*see* **Note 62**), heat at 65 °C for 15 min then chill on ice for 2 min.

4. Using a suitable pipettor and pipette tip (one with a thin tip, to allow loading of the 0.75 mm gel; *see* **Note 63**), load the sample into a lane of a sequencing gel (Subheading 3.2.5). Run the gel and process as described in Subheading 3.2.5. (*See* **Note 59**.)

---

4 Notes

1. A wide range of vectors and systems for expression of affinity-tagged fusion proteins in *E. coli* are available. In the authors' laboratories, the affinity tags have included maltose binding protein (MBP), a novel calmodulin-binding domain (present in the pCAL series of vectors, presently available from Agilent Technologies), glutathione-S-transferase (GST), and a polyhistidine tag (HH). These plasmids may possess different antibiotic resistance determinants, which may permit the occupation of multiple different plasmids in a cell. Regardless, the choice of antibiotic resistance marker is one of the considerations in the design and execution of the experiment. Depending on the protein of interest, both amino and carboxy terminal tags may need to be tested empirically for their effects on protein–protein and/or RNA–protein interactions. Typically, expression of the fusion protein and localization of it to the cytosol is successful. If problems are encountered with insoluble or degraded protein, targeting of the expressed protein to the periplasmic space might be advantageous, which may be afforded by systems from some vendors.

2. These are the two strains that have been used to best effect in the authors' laboratories. However, there is a wide range of specialized *E. coli* strains suitable for the production of recombinant proteins. For example, if no protein is expressed or low-levels that are unsuitable for the experimental needs are

encountered, a strain coding for rare tRNAs (such as BL21(DE3) pLysS) might be beneficial for protein expression. Moreover, in cases where formation of disulfide bonds between key cysteine residues is an issue, *E. coli* strains such as Origami™ (DE3) that enhance disulfide bond formation in the cytoplasm can be considered. One often overlooked consideration that should be kept in mind is the fact that *E. coli* BL21-related strains are not *E. coli* K12 derivatives, but rather are derived from *E. coli* B. Thus, research carried out with these strains is not automatically exempt from the NIH Guidelines for Research Involving Recombinant or Synthetic Nucleic Acid Molecules (NIH Guidelines), and must be duly registered with the Institutional Biosafety Committee or analogous administrative unit.

3. 2 % glucose is added to the medium when growing cells for protein production to repress the expression of native *E. coli* amylase genes; *E. coli* amylase can degrade the amylose resin used to purify MBP fusion proteins. Also, adding 1 % glycerol to the media increases the available carbon for the bacterial cells to channel into protein production.

4. If protein degradation is a problem, include a protease inhibitor cocktail (exclude EDTA from cocktail when using metal affinity purification because this will strip Ni^{2+} or Co^{2+} from the column). Also, for some proteins, it may be desirable to include dithiothreitol (DTT) in the purification, wash, and elution buffers. This is not routinely done for plant poly(A) factor purification as DTT inhibits the activity of the Arabidopsis CPSF30 protein [18].

5. Use freshly transformed bacterial cells for protein expression because protein expression levels and enzyme activity can decrease over time despite being maintained on selection.

6. At the outset of a project, it is advisable to examine several independent transformants for expression and protein quality. This assures that at least one culture with good protein expression levels will be identified.

7. For initial isolation of clones, antibiotic concentrations will be as recommended by the suppliers of the different expression vectors. However, subsequent steps (as for protein production) may entail the use of lower antibiotic concentrations, since this may be less metabolically onerous for the bacterial cells allowing more energy to be directed towards protein expression.

8. For optimal aeration of the culture, only 20–25 % of the flask volume should be occupied with liquid culture.

9. Growth at 23–25 °C increases protein yields and is particularly suited for the expression of proteins that are inclined to form insoluble inclusions.

10. Cells may be stored at −80 °C for several days before subsequent use. However, longer-term storage is not advised.

11. This step constitutes a preliminary characterization of the fusion proteins, and is done to screen independent recombinants for efficient production of full-sized, un-degraded proteins.

12. This generic protocol has been used with fusion proteins that carry hexahistidine (HH), maltose-binding protein (MBP), glutathione-S-transferase (GST), and calmodulin-binding domain (CBD) tags [8–14, 18, 19]. For all but HH fusion proteins, the base purification buffer system (Subheading 2.1, **item 9**) may be used. Note that CBD fusion proteins require the inclusion of 5 mM calcium in the purification buffer. The HH buffer system has a different base buffer (sodium phosphate), and requires different concentrations of imidazole in the wash and elution solutions. These are listed in Subheading 2.1, **items 19–24**.

13. Downstream applications using HH-tagged proteins require additional precautions at this point. Thus, if protease inhibitors are included in the purification buffer used to resuspend cells, care must be taken to use inhibitors that are compatible with the His-Select affinity matrix. In particular EDTA used to inhibit metalloproteases can strip the affinity gel of cobalt. Including low concentrations of imidazole in the purification and wash buffers can help to prevent co-purification of undesirable proteins. Test with low concentrations (5–20 mM) first, and then increase if necessary.

14. Sonication can damage hearing therefore adequate ear protection should be worn at all times. Additionally, sonication generates considerable amounts of heat in the sample when the sounds waves are being generated. Thus, sonication should be conducted in a cold room with the sample on ice if possible. Practice beforehand can be beneficial to prevent foaming of the sample, which will decrease the efficiency of cell lysing if foam is generated.

15. Be sure to use the appropriate tubes for centrifugation. Plastic tubes might split open at high centrifugal forces whereas certain glass tubes might disintegrate.

16. As described here, agarose or Sepharose-based affinity matrices with immobilized cations (cobalt, for HH fusion proteins), amylose (for MBP fusions), glutathione (for GST fusions), or calmodulin (for CBD fusions) may be used.

17. Fusion proteins containing the MBP, GST, and CBD tags may all be processed with the same base buffer system (Subheading 2.1, reagents 9 and 10). HH-containing fusion proteins should be processed using a different buffer system (Subheading 2.1, reagents 19–24).

18. To minimize disturbing the bedded resin/gel, allow the lysate to slowly and gently run down the side of the column until ~1–2 mL has been applied to the column bed. After this, the remaining lysate may be applied more rapidly, but still gently enough to avoid disturbance of the resin bed.

19. The flow rate can be estimated using a tube with volume marks and a timer during the equilibration step of column preparation.

20. The buffer compositions given in Subheading 2.1 are those generally used over the years by the authors' laboratories. However, in a more general sense, the column wash buffers contain high NaCl concentrations (1–2 M), and are intended to remove proteins that bind nonspecifically to the affinity matrix or to the purified protein itself. The purification and elution buffers have more modest NaCl concentrations (between 100 and 200 mM) and are compatible with subsequent assays. The washes may be discarded or saved, as needed for subsequent analyses.

21. The choice of eluent (maltose, glutathione, or EGTA, for MBP, GST, and CBD fusions, respectively), is determined by the affinity tag. For MBP, GST, and CBD fusions, the eluent is added to the base purification buffer (Subheading 2.1, **item 9**), usually at concentrations of 10 mM. The elution buffer for HH fusions is different, and described in Subheading 2.1, **item 24**.

22. If greater than 100 mM imidazole concentrations are needed to elute the protein from the column, dilute the final eluate to less than 100 mM imidazole with HH purification buffer that does not contain imidazole. Concentrations of imidazole greater than 100 mM are not compatible with the Amicon Ultra centrifugal filter devices used for buffer exchange and concentration.

23. The time needed to reach this volume will vary with the sample, but will typically be between 10 and 30 min.

24. Short-term storage may be done at –20 °C, while longer-term storage should be done at –80 °C. In either case, before storage, a sample should be assessed by SDS-PAGE followed by staining, as described in Subheadings 3.2.1 and 3.2.2.

25. It may be desirable to further process purified proteins by removal of the respective affinity tags. The cloning vectors associated with each system allow for this possibility, by incorporating recognition sites for site-specific proteases into the respective constructs. In our experience, this is not necessary for most of the characterizations that have been described. Indeed, for at least one plant polyadenylation factor subunit, proteolytic removal of the (MBP) tag results in precipitation of the polyadenylation factor subunit. More generally, affinity tags are known to promote solubilization of fusion proteins

Table 1
Recipes for SDS-polyacrylamide gels

Solution[a]	Separating gels[b]			Stacking[c]
Gel (%)	10	12.5	15	NA
SDS-PAGE gel buffer (mL)	25	25	25	5.0
10 % SDS (mL)	0.5	0.5	0.5	0.1
30 % acrylamide solution (mL)	16.7	20.8	25	1.0
Water (mL)	9.5	5.4	1.2	3.9
25 % APS (µL)	80	80	80	35
TEMED (µL)	40	40	40	8

[a]*See* Subheading 2.1 for descriptions of these stock solutions
[b]These volumes will produce sufficient solution to pour four 8 cm × 10 cm separating gels of 1.5 mm thickness in GE HealthCare Life Sciences SE250 and SE260 Mini-Vertical Electrophoresis units
[c]These volumes will produce sufficient solution to pour four 2 cm × 10 cm stacking gels of 1.5 mm thickness in GE HealthCare Life Sciences SE250 and SE260 Mini-Vertical Electrophoresis units

[20]; for this reason, removal of the affinity tags is usually not done in our laboratories.

26. This recipe results in a separating gel that is 10 % acrylamide, and is sufficient for one 10 × 10 cm gel of 1.5 mm thickness. This can be varied (lower percentage for large proteins, higher percentage for small polypeptides) by changing the quantities of water and acrylamide that are used. Table 1 presents recipes for the more common separating gel concentrations used in our laboratories.

27. A number of approaches may be used to assemble a leak-free gel-casting sandwich. These include the use of manufactured systems that incorporate sealing mechanisms, or the use of "home-made" apparatuses built around custom-cut glass plates and spacers, and sealants such as agarose that are applied once a sandwich is assembled. The recipes described here are suited for use with the GE HealthCare Life Sciences SE250 and SE260 Mini-Vertical Electrophoresis units; these allow for gels of 10 × 10 cm in size to be cast and used. SDS and native gels are usually 1.5 mm in thickness, and sequencing gels are 0.75 mm thick.

28. Attach one end of a thick-walled hose to a Pasteur pipette that has been inserted through the middle of a rubber stopper. Attach the other end of the hose to a vacuum line then place the rubber stop on top of the flask. Slowly open the valve of the vacuum line making sure not to allow the liquid to suck into the vacuum line. After ~10–20 min, close the vacuum line valve

and allow the pressure to equalize in the flask. Do not pull the rubber stopper off or the liquid may erupt from the flask.

29. The stacking gel should polymerize within 30 min. Once polymerized, if the gel is not going to be used immediately, it may be wrapped in water soaked paper towels and plastic wrap, and then stored at 4 °C for up to 1 week.

30. Include a protein-sizing standard on one side of the gel. By loading size standards on one edge of the gel, an asymmetry is introduced. This asymmetry helps to follow the samples after further manipulations that may include gel inversions owing to the transfer of samples to nitrocellulose membranes.

31. Lids of 1 mL pipette tip boxes work well as plastic trays. Liberally apply the Coomassie stain to make sure the gel is fully submerged with no portion of the gel exposed. This helps prevent making a mess during the microwave step. A dedicated microwave in a fume hood should be used because of the vapors from the acetic acid. If this is not possible, be sure to leave the microwave door open afterwards to allow it to aerate.

32. Coomassie staining solution can be reused several times.

33. The sponge or low-lint wipe serves to remove the stain from the solution and speeds the destaining process.

34. The transfer procedure described here has been used with tank-based electro-transfer apparatuses such as the Bio-Rad Mini Transblot Electrophoretic Transfer Cell. Other systems will require different arrangements, buffers, and quantities, but almost all will retain the basic features of the methods described here—namely, direct contact of the gel with a nitrocellulose membrane, followed by electrophoretic transfer of proteins from the gel to the membrane.

35. Construct the transfer sandwich in a tray filled with transfer buffer. Keeping the sandwich submerged in liquid and rolling with a glass rod or plastic pipette prevents air bubbles from remaining in the sandwich. After separating the glass plates, leave the gel stuck to one of the plates then place a piece of 3 mm Whatman paper to the gel. Peel the gel from the plate and place it on top of the first sponge followed by the other components.

36. It is very important to disconnect all power sources. Full-tank apparatuses such as those used in the authors' laboratories utilize very high amperages for the transfer and require exacting and scrupulous attention to prevent accidental exposure of personnel to these dangerous currents.

37. The volumes of TTBS and Blocking Solution will be determined by the sizes of the trays used for washes. Regardless, enough solution should be used to completely immerse the membrane once it is placed in the tray.

38. The primary antibody will be specific for the affinity tag that is being detected—typically, HH, MBP, or GST. The antibody quantity may be estimated from the datasheets provided by the supplier; typically, a 1:1,000 to 1:10,000 dilution will be sufficient. However, empirical tittering of the antibody may be needed occasionally.

39. Fusion proteins containing the CBD are usually detected by incubating filters with biotinylated calmodulin. This reagent is supplied with the vectors used (presently, available from Agilent Technologies). Thus, for Subheading 3.2.3, **step 5**, the 10 mL of Blocking Solution is replaced with 1 mL of 1× TTBS + 1 mM $CaCl_2$. The filter is wrapped in plastic wrap to prevent evaporation and incubated at 4 °C overnight.

40. The secondary antibody will be one that recognizes the primary antibody, and to which alkaline phosphatase has been conjugated. The specific choice will depend on the nature of the primary antibody used in Subheading 3.2.3, **step 4**; thus, a mouse monoclonal primary antibody will dictate that the secondary antibody be an alkaline phosphatase-conjugated rabbit anti-mouse antibody. Typically, the secondary antibody may be used at a 1:10,000 dilution, although some empirical testing may be required.

41. For the detection of CBD fusion proteins, no secondary antibody is required. Instead, streptavidin that has been conjugated with alkaline phosphatase is used, according to the supplier's recommendations. In this case, at Subheading 3.2.3, **step 6**, the 10 mL of new Blocking Solution is replaced with 10 mL of 1× TTBS + 1 mM $CaCl_2$.

42. For the detection of CBD fusion proteins, the 1× TTBS wash solution is replaced with 1× TTBS + 1 mM $CaCl_2$.

43. The choice to use reagents centered on alkaline phosphatase reflects the fact that the protein quantities that are being dealt with are high enough to preclude the use of more sensitive (and expensive) reagents (such as chemiluminescence kits). However, the latter may be used by replacing the alkaline phosphatase-conjugated reagents with horse radish peroxidase-conjugated antibodies and streptavidin.

44. The recipe given here yields a gel that is 4 % in acrylamide and 0.08 % in N,N'-methylenebisacrylamide. These concentrations have been used routinely to measure RNA binding by plant CPSF30, Fip1, and CstF77 proteins, and provide good resolution of complexes from free RNA when used in gels that are 1.5 mm in thickness and 8 or 10 cm in length. However, it may prove desirable to vary these concentrations; if so, empirical studies are recommended to identify gel concentrations suited for the particular application.

Table 2
Recipes for sequencing gels

Gel percent	7.5 %		10 %		15 %		20 %	
# of gels[a]	1	2	1	2	1	2	1	2
Urea (g)	14.7	21	14.7	21	14.7	21	14.7	21
Acrylamide (g)	2.7	3.9	3.5	5	5.3	7.6	7	10
Bisacrylamide (mg)	78	85	98	140	157	224	196	280
AP (mg)	21	30	21	30	21	30	21	30
H_2O (mL)	5	7	5	7	5	7	5	7
5× TBE (mL)	7	10	7	10	7	10	7	10
Bring to vol. (mL)	35	50	35	50	35	50	35	50
TEMED (µL)	14	20	14	20	14	20	14	20

[a]Number of gels that can be poured with the indicated quantities of components

45. Non-denaturing polyacrylamide gels will typically polymerize within 30 min. Polymerized gel sandwiches can be wrapped in wet paper towels and plastic wrap and stored at 4 °C for up to 1 week.

46. It is preferable to conduct the electrophoretic separation at a high enough voltage to complete the separation within 30–60 min; shorter times minimize the in situ dissociation of RNA–protein complexes and the problems of interpretation that are associated with this. Accordingly, electrophoresis is conducted at 4 °C to avoid gel heating effects associated with elevated currents and voltages. Under such conditions, running the tracking dye to the bottom of a 10 cm gel will result in running a moderately sized RNA at least 75 % of the way down the gel.

47. These volumes are sufficient to pour a single 10×10 cm gel of 0.75 mm thickness. *See* Table 2 for more common recipes that make one or two gels of varying acrylamide concentrations.

48. It is very important to cool the gel solution to room temperature before adding TEMED. Polymerization will proceed very rapidly if the solution is at elevated temperatures, so fast that the mixture may polymerize before it can be poured into a gel sandwich.

49. These sequencing gels are cast in set-ups for 8 or 10 cm in length, 10 cm in width, and 0.75 mm in thickness.

50. For large affinity tags such as MBP or GST, the unmodified versions will suffice as negative controls. For smaller tags (HH, CBD), tagged forms of GFP or some other well-behaved marker protein should be used.

51. Other buffer components and reagents may be added to these reactions as needed. Note that the purification buffer used to suspend the cell pellets (Subheading 3.1.2, **step 1**) will be determined by the choice of "bait" in the co-purification assays. Thus, if the "bait" will be the HH-tagged protein, both lysates should be prepared in the HH-tagged protein purification buffer (Subheading 2.1, **item 22**). Conversely, for CBD-, MBP-, or GST-tagged "baits," the general purification buffer (Subheading 2.1, **item 9**) should be used, even when making lysates containing HH-tagged proteins. Protein concentrations may be adjusted by replacing the respective crude lysate with the corresponding purification buffer, as deemed necessary for the particular experiment.

52. Incubation times should be empirically determined. For some fusion proteins, lengthy incubation times may lead to increased extents of protein denaturation, which in turn may lead to non-specific aggregation and adsorption to the affinity matrices.

53. Assessment of immunoblots may be qualitative (plus-or-minus) or more quantitative. Reliable quantitation requires concomitant analysis of control samples and of standards (usually, known quantities of the various fusion proteins). Practices such as titration of standard samples (typically, at least two orders of magnitude separated and analyzed on the same gel) are very important, as it is otherwise very difficult to evaluate differences in band intensity.

54. The radiolabeled and unlabeled RNAs will usually be synthesized using any of a number of commercially available in vitro transcription kits. RNA synthesis using these is conducted exactly according to manufacturers' recommendations and is not described in detail here. These RNAs will typically be derived from parts of plant transcription units that include a polyadenylation signal; given the nature of the plant polyadenylation signal, these RNAs may be as large as 100–150 nucleotides [21]. Regardless of this, the specific activities of the radiolabeled RNAs should be on the order of 10^5–10^6 dpm/µg, such that between 10^5 and 10^6 dpm are added per reaction. This quantity of RNA should constitute no more than 10 % of the total volume of the binding reaction. To this end, it may be required to dilute a more concentrated stock of labeled RNA with the appropriate buffer that is used for the binding reaction.

55. Unlabeled RNA is added as a competitive inhibitor, or as a nonspecific carrier that may interfere with nuclease contamination or nonspecific RNA–protein interactions. Specific inhibitors for RNA-binding reactions will typically be derived from in vitro transcription using any of a number of commercially available in vitro transcription kits. These will typically be added in 1- to 100-fold molar excess over the radiolabeled

RNA substrate. Nonspecific inhibitors may be commercial preparations of total RNA or tRNA from a bacterial or yeast source; quantities used in assays may be as great as 1–10 µg per reaction, and should be empirically determined for each particular RNA–protein interaction. In all cases, these RNAs should be added from stocks prepared with the respective buffer for the binding reactions; given that these will include divalent metal cations, these stocks should be freshly prepared so as to minimize RNA degradation during storage.

56. This is a typical range of protein quantities, and may be as much as 4–5 µg of the purified protein. If the experiment demands comparable total protein concentrations across a wide range of quantities of a particular polyadenylation factor subunit, the balance may be made up with bovine serum albumin.

57. In most cases, the respective protein purification buffer (NOT the storage buffer) will be most appropriate for RNA-binding reactions. However, the concentrations of some components of these buffers, such as Mg^{2+} and NaCl, may need to be varied so as to determine optimal binding conditions. Also, inclusion of compounds such as heparin sulfate can minimize nonspecific binding of RNA by the target protein. Finally, an appropriate quantity of commercially available RNase inhibitor (up to 10 U of the human placenta RNase inhibitor from New England Biolabs or an equivalent activity of a comparable inhibitor) may be needed to inhibit trace amounts of ribonuclease contamination.

58. Optimal reaction times and temperatures should be determined empirically. Typically, for plant polyadenylation factors, temperatures of 25 30 °C are optimal. Times will be impacted in large part by the quantities and inherent activities of the proteins of interest.

59. The primary experimental output for the assays described in Subheadings 3.3.2 and 3.3.3 are images or autoradiographs that are based on a common set of electrophoretic separations. Autoradiographs may be analyzed using any of a number of software packages. ImageJ is a freely available tool that is widely used for such purposes and is the package used by the authors. Results may be presented in terms of absolute quantities of radioactivity in bands of interest. This requires simultaneous separation, detection, and analysis of known quantities (molar as well as radioactivities) of substrate RNAs. More commonly, results are presented in terms of fractional, normalized abundance, where all the radioactive signals in a lane are totaled and the fractional abundance of radioactivity in specific bands in the lane are calculated. Biochemical parameters such as Km's, equilibrium constants, and inhibition constants may be determined using these assays by measuring fractional abundance as

a function of the concentrations of substrate, enzyme, or inhibitor. These matters have been discussed extensively in the literature (e.g., [22, 23]).

60. In most cases, the respective protein purification buffer (NOT the storage buffer) will be most appropriate for processing reactions. However, the concentrations of some components of these buffers, such as Mg^{2+} and NaCl, may need to be varied so as to determine optimal activities. Other compounds may be called for, depending on the particular activity being assayed. For example, poly(A) polymerase activity will require ATP (in the 50–500 μM concentration range), as may some RNA processing reactions. Other nucleotidyltransferase activities will require similar concentrations of other NTPs. Nuclease reactions may require divalent metal cations (*see* [8, 14, 18, 19] for examples where different Arabidopsis polyadenylation-related proteins were assayed using this general method).

61. As with EMSA reactions, optimal reaction times and temperatures for RNA processing and modification reactions should be determined empirically. Typically, for plant polyadenylation factors, temperatures of 25–30 °C are optimal. Times will be impacted in large part by the quantities and inherent activities of the proteins of interest. Time courses should be designed so that no more than 90 % of the input RNA is broken down after the longest reaction time; for precise kinetic analysis, most of the substrate should remain un-degraded or unprocessed.

62. In the assay described here, no attempt is made to recover and purify the products of the RNase reactions. The traces of organic solvents that carry through into the aqueous phase of the reactions is minimal and does not interfere with the electrophoretic separation of the nucleic acids. However, the buffer components will affect electrophoretic behavior; thus, it is very important to use equal volumes of sample, all of which have identical buffer compositions.

63. To facilitate the gel loading process, capillary tips can be used which make getting the sample into the well a bit easier. If well overflow is an issue, load less sample volume per well.

Acknowledgements

The authors thank Dr. Arthur G. Hunt at the University of Kentucky, Lexington, KY, USA for helpful comments and editorial guidance with this manuscript. Dr. Addepalli was supported in part by grant MCB-0313472 from the US National Science Foundation (awarded to Drs. A. G. Hunt and Q. Q. Li).

References

1. Edmonds M (2002) A history of poly A sequences: from formation to factors to function. Prog Nucleic Acid Res Mol Biol 71:285–389

2. Millevoi S, Vagner S (2010) Molecular mechanisms of eukaryotic pre-mRNA 3′ end processing regulation. Nucleic Acids Res 38(9): 2757–2774. doi:10.1093/nar/gkp1176

3. Moore MJ, Proudfoot NJ (2009) Pre-mRNA processing reaches back to transcription and ahead to translation. Cell 136(4):688–700. doi:10.1016/j.cell.2009.02.001

4. Chan S, Choi EA, Shi Y (2011) Pre-mRNA 3′-end processing complex assembly and function. Wiley Interdiscip Rev RNA 2(3):321–335. doi:10.1002/wrna.54

5. Ryan K, Bauer DL (2008) Finishing touches: post-translational modification of protein factors involved in mammalian pre-mRNA 3′ end formation. Int J Biochem Cell Biol 40(11):2384–2396. doi:10.1016/j.biocel.2008.03.016

6. Shi Y, Chan S, Martinez-Santibanez G (2009) An up-close look at the pre-mRNA 3′-end processing complex. RNA Biol 6(5):522–525

7. Zhao H, Zheng J, Li QQ (2011) A novel plant in vitro assay system for pre-mRNA cleavage during 3′-end formation. Plant Physiol 157(3):1546–1554. doi:10.1104/pp. 111.179465

8. Addepalli B, Hunt AG (2007) A novel endonuclease activity associated with the Arabidopsis ortholog of the 30-kDa subunit of cleavage and polyadenylation specificity factor Nucleic Acids Res 35(13):4453–4463. doi:10.1093/nar/gkm457

9. Addepalli B, Hunt AG (2008) The interaction between two Arabidopsis polyadenylation factor subunits involves an evolutionarily-conserved motif and has implications for the assembly and function of the polyadenylation complex. Protein Pept Lett 15(1):76–88

10. Addepalli B, Limbach PA, Hunt AG (2010) A disulfide linkage in a CCCH zinc finger motif of an Arabidopsis CPSF30 ortholog. FEBS Lett 584(21):4408–4412. doi:10.1016/j.febslet.2010.09.043

11. Bell SA, Hunt AG (2010) The Arabidopsis ortholog of the 77 kDa subunit of the cleavage stimulatory factor (AtCstF-77) involved in mRNA polyadenylation is an RNA-binding protein. FEBS Lett 584(8):1449–1454. doi:10.1016/j.febslet.2010.03.007

12. Delaney KJ, Xu R, Zhang J, Li QQ, Yun KY, Falcone DL, Hunt AG (2006) Calmodulin interacts with and regulates the RNA-binding activity of an Arabidopsis polyadenylation factor subunit. Plant Physiol 140(4):1507–1521. doi:10.1104/pp. 105.070672

13. Elliott BJ, Dattaroy T, Meeks-Midkiff LR, Forbes KP, Hunt AG (2003) An interaction between an Arabidopsis poly(A) polymerase and a homologue of the 100 kDa subunit of CPSF. Plant Mol Biol 51(3):373–384

14. Forbes KP, Addepalli B, Hunt AG (2006) An Arabidopsis Fip1 homolog interacts with RNA and provides conceptual links with a number of other polyadenylation factor subunits. J Biol Chem 281(1):176–186. doi:10.1074/jbc. M510964200

15. Hornyik C, Terzi LC, Simpson GG (2010) The spen family protein FPA controls alternative cleavage and polyadenylation of RNA. Dev Cell 18(2):203–213. doi:10.1016/j.devcel. 2009.12.009

16. Simpson GG, Dijkwel PP, Quesada V, Henderson I, Dean C (2003) FY is an RNA 3′ end-processing factor that interacts with FCA to control the Arabidopsis floral transition. Cell 113(6):777–787

17. Bertani G (2004) Lysogeny at mid-twentieth century: P1, P2, and other experimental systems. J Bacteriol 186(3):595–600

18. Addepalli B, Hunt AG (2008) Redox and heavy metal effects on the biochemical activities of an Arabidopsis polyadenylation factor subunit. Arch Biochem Biophys 473(1):88–95. doi:10.1016/j.abb.2008.02.027

19. Addepalli B, Meeks LR, Forbes KP, Hunt AG (2004) Novel alternative splicing of mRNAs encoding poly(A) polymerases in Arabidopsis. Biochim Biophys Acta 1679(2):117–128. doi:10.1016/j.bbaexp.2004.06.001

20. Kapust RB, Waugh DS (1999) Escherichia coli maltose-binding protein is uncommonly effective at promoting the solubility of polypeptides to which it is fused. Protein Sci 8(8):1668–1674. doi:10.1110/ps.8.8.1668

21. Hunt AG (2008) Messenger RNA 3′ end formation in plants. Curr Top Microbiol Immunol 326:151–177

22. Fried MG (1989) Measurement of protein-DNA interaction parameters by electrophoresis mobility shift assay. Electrophoresis 10(5–6):366–376. doi:10.1002/elps.1150100515

23. Hellman LM, Fried MG (2007) Electrophoretic mobility shift assay (EMSA) for detecting protein-nucleic acid interactions. Nat Protoc 2(8):1849–1861. doi:10.1038/nprot.2007.249

Chapter 10

Detection of Disulfide Linkage by Chemical Derivatization and Mass Spectrometry

Balasubrahmanyam Addepalli

Abstract

The location of disulfide linkage(s) or status of unpaired cysteines is a critical structural feature required for the characterization of three-dimensional structure of a protein and for the correlation of protein structure–function relationships. Cysteine, with its reactive thiol group, can undergo enzymatic or oxidative posttranslational modification in response to changing redox conditions to signal a cascade of downstream reactions. In such a situation, it becomes even more critical to obtain the information on the pair of cysteines involved in such a redox switch operation. Here, a method involving chemical derivatization and liquid chromatography–mass spectrometry (LC-MS) is described to determine the cysteine residues involved in disulfide bond formation for a protein containing multiple cysteines in its sequence.

Key words Disulfide linkage, Mass spectrometry, Posttranslational modification

1 Introduction

Polyadenylation is an essential posttranscriptional processing event of eukaryotic messenger RNA. It involves complex interplay of cis-acting RNA elements and trans-acting protein factors [1–4]. Among these protein factors, saving one homolog of CstF64 in Arabidopsis (AT1G73840), rest of the polyadenylation protein factors have cysteine residues ranging from 0.3 to 4 for every 100 amino acids (www.arabidopsis.org). CPSF30 has the maximum number of cysteines (4 for 100 amino acid residues) followed by CPSF73II with 3.4 residues. While the cysteines in CPSF30 are predicted to be part of CCCH zinc fingers, such an organization is not postulated for CPSF73II subunit. Proteins with multiple cysteine residues have the potential to play an important role in redox stress responses because of their ability to oxidize. Indeed, the 30 kDa subunit of Cleavage and Polyadenylation Specificity Factor has been implicated in oxidative stress signaling in *Arabidopsis thaliana* [5].

Arthur G. Hunt and Qingshun Quinn Li (eds.), *Polyadenylation in Plants: Methods and Protocols*, Methods in Molecular Biology, vol. 1255, DOI 10.1007/978-1-4939-2175-1_10, © Springer Science+Business Media New York 2015

Enzymatic or oxidative disulfide bond formation is one of the key posttranslational cysteine modification that affects structure and function of a protein. The reactive thiol side chain of cysteine is vulnerable to modification in response to changing redox conditions of the cell; such modification can provide an opportunity for pair/s of cysteine residues to work as redox switches [6, 7]. Here, the side chain of cys can function as a sensor or switch through its reactive thiol group, flipping between the reduced and oxidized state in response to fluctuations in the reactive oxygen species or nitrogen species (RO/NS) [6, 8, 9]. Changes in redox environment can form or break a localized disulfide bond, thereby introducing a structural change in the conformation of a molecule and changing its function. Thus, identification of the status of these critical residues is a first step in the comprehensive characterization of the redox switch and for structure–function correlation.

A variety of methods have been introduced to investigate disulfide linkage and status of unpaired cysteine residues [10]. In general, proteomic approaches have been playing a major role in identification of individual Cys residues involved in thiol modification [11–14]. Assignment of disulfide bonds, however, is not trivial and can still present several challenges [15, 16]. For example, disulfide bond scrambling (exchange of partners between thiols and disulfides) can occur in gas phase during mass spectrometry (MS) analysis [16]. In solution-phase, such scrambling occurs when the pH is too high (pH > 8), so it is advisable to perform procedures near neutral conditions [17]. Scrambling will be more problematic with bottom-up (peptides to protein) proteomic approaches, if a protein contains multiple cysteine residues such as those in CCCC or CCCH-type zinc finger proteins [18, 19], thus leading to erroneous data and interpretation. Hence, it becomes all the more important to block the free thiol groups in such proteins before digestion with a protease. In the present method, the protein is alkylated before or after treating with reducing agent, but before digestion so that the disulfide linkage is uniquely identified. The method is compatible with the LC-MS (liquid chromatography coupled with mass spectrometry) analysis.

2   Materials

All solutions are required to be prepared with autoclaved ultrapure water (deionized water with a sensitivity of 18 MΩ cm at 25 °C) and analytical grade reagents. Store the reagents as indicated. Conform to all waste disposal and safety regulations while preparing and disposing the materials.

1. Denaturing and digestion buffer solution: 0.1 M Tris–HCl (pH 7.5), 10 mM $CaCl_2$, 8 M urea (*see* **Note 1**).

2. Reducing agent: 45 mM DTT, store at –20 °C.

3. Alkylating reagent: 100 mM Iodoacetamide, prepare fresh.

4. Cold acetone.

5. ProteaseMAX™ Surfactant, Trypsin enhancer, store at –20 °C (Promega; *see* **Note 2**).

6. Trypsin Gold, Mass Spectrometry grade (Promega).

7. 50 mM ammonium bicarbonate (pH ~7.6).

8. Heating block.

9. C18 tips such as ZipTips™ (Millipore).

10. 2.5 % TFA (trifluoroacetic acid).

11. Wetting solution: 50 % acetonitrile 50 % Water.

12. Equilibration solution: 0.1 % TFA.

13. Washing solution: 0.1 % TFA, 5 % methanol.

14. Elution solution: 50 % acetonitrile, 0.1 % TFA.

15. Mobile phase A for reverse phase high performance liquid chromatography (RP-HPLC): 95 % water, 5 % acetonitrile, 0.1 % formic acid.

16. Mobile phase B for reverse phase high performance liquid chromatography (RP-HPLC): 95 % acetonitrile, 5 % water, 0.1 % formic acid.

17. Biobasic C18 packed tips: 0.075×100 mm tips or 1×100 mm C18 X-bridge column (Waters).

3 Methods

3.1 Description of Experimental Rationale

Iodoacetamide treatment results in the addition of carbamido-methyl (alkyl) group (+57 Da) to any free thiol group. Cysteines associated with disulfide linkage do not react with iodoacetamide, and therefore, are not alkylated. However, DTT treatment cleaves the disulfide bond and the resulting cysteines are amenable to carbamidomethylation thus adding a mass of 58 Da for each cysteine. Characterization of proteins by mass spectrometry could be used to study the existence and chemical nature of disulfide bonds using this susceptibility to iodoacetamide. Specifically, a comparison of two data sets (i.e., with and without DTT-treatment) would reveal the presence or absence of a disulfide bond for a given cysteine pair, and would identify the specific amino acid residues engaged in such bonds.

Briefly, the workflow for these determinations is as follows: preparation of a protein sample, parallel treatments of reduced and unreduced samples with iodoacetamide, digestion of proteins with trypsin, and analysis of peptides by Liquid Chromatography coupled with Mass Spectrometry (LC-MS).

3.2 Sample Preparation

In the following, carry out all procedures at the specified temperature. The experimental steps are performed at room temperature if temperature is not specified.

3.2.1 Protein Precipitation

1. Precipitate protein with 70 % cold acetone (for 30 μl protein solution, add 70 μl of cold acetone) for at least 3 h or overnight at –20 °C (*see* **Note 3**).

2. Centrifuge the suspension at 12,000×*g* for 15 min and discard the supernatant. Add 1 ml of 70 % acetone to the pellet, vortex briefly, centrifuge at 12,000×*g* for 7 min, and discard the supernatant. Repeat the process one more time.

3. Two equal amounts of protein are processed in parallel as stated below in **step 2** of the following section.

3.2.2 Denaturation and Reduction of Protein

1. Resuspend the pellet in denaturation and digestion buffer so as to obtain a concentration of 1 μg/μl (typically, between 15 and 30 μl; *see* **Note 4**).

2. Add DTT to a final concentration of 22 mM to one protein sample and label it as *reduced*. Add water to the second sample, and label it as *unreduced*.

3. Incubate both mixtures at 60–70 °C for 20–45 min (*see* **Note 5**) on a heating block.

4. Bring the samples to room temperature and add iodoacetamide to 55 mM final concentration. Incubate in the dark for 30 min.

5. Dilute each sample with water so as to obtain a urea concentration of less than 1 M, and digest with trypsin (at a weight/weight ratio of 1:25 for Trypsin–protein) overnight at 37 °C.

6. Add trifluoroacetic acid (TFA) to 0.1 % final concentration to stop the digestion.

3.2.3 Purification of Peptides Using C18 Ziptips

The peptides resulting from trypsin digestion are purified using ZipTip-based purification. The maximum binding of peptides to the ZipTip is achieved in the presence of TFA or other ion-pairing agents. The final TFA concentration should be between 0.1 and 1.0 % at a pH of <4 (*see* **Note 6**).

Equilibrate the ZipTip for Sample Binding:

1. Prewet the tip by depressing pipettor plunger to a dead stop using the maximum volume setting of 10 μl. Aspirate wetting solution (50 % acetonitrile 50 % Water) into tip. Dispense to waste. Repeat.

2. Equilibrate the tip for binding by washing with the equilibration solution (0.1 % TFA) three times.

Bind and Wash the Peptides:

3. After equilibration, bind peptides to ZipTip by fully depressing the pipettor plunger to a dead stop. Aspirate and dispense sample 3–7 cycles for simple mixtures and up to 10 cycles for maximum binding of complex mixtures.

4. Wash tip and dispense to waste using at least 5 cycles of wash solution. A 5 % methanol in 0.1 % TFA/water wash can improve desalting efficiency.

Peptide Elution:

5. Carefully aspirate and dispense 5 μl elution solution (50 % acetonitrile, 0.1 % TFA) through ZipTip at least three times without introducing air (*see* **Note 7**).

6. Dry the eluted peptides in a SpeedVac fitted with a cold trap. Dried peptides may be stored or used immediately for LC-MS analysis.

3.3 Liquid Chromatography Coupled with Mass Spectrometry (LC-MS)

*3.3.1 Reverse Phase High Performance Liquid Chromatography (RP-HPLC) (See **Note 8**)*

The peptide digests from Subheading 3.2.3, **step 6** are separated either on Xbridge C18 1.0 × 150 mm column (Waters) or on nanospray tips (Biobasic C18 packed tips: 0.075 × 100 mm tips) at room temperature.

For both columns, the following gradient scheme is used:

1. Load the sample at 95 % Mobile phase A (95 % water, 5 % acetonitrile, 0.1 % formic acid) 5 % Mobile phase B (95 % acetonitrile, 5 % water, 0.1 % formic acid) for 5 min.

2. Run a linear gradient from 95 % A/5 % B to 50 % A/50 % B in 55 min.

3. Run a linear gradient from 50 % A/50 % B to 20 % A/80 % B in 3 min.

4. Continue with an isocratic step at 20 % A/80 % B for another 12 min.

5. Re-equilibrate the column with at least 10 column volumes of initial mobile phase composition (95 % A and 5 % B).

The flow rates are 70 μl/min (for Xbridge column) and 0.5 μl/min (nanospray tips) with the gradient remaining same for both columns. Eluted peptides are suitable for either electrospray or nanospray ionization, respectively, for transferring peptides from liquid phase to gas phase.

3.3.2 Mass Spectrometry

Conditions for ionization and detection depend on the type of mass spectrometer used, and specific details will vary based on the instrument, operator, and desired outcomes. However, some general guidelines can be spelled out. Recommendations based on past experience (e.g., [19]) are outlined in this section.

A mass spectrometer can have one (or more) ionization sources and one (or more) mass analyzers. However, electrospray ionization (ESI) source is mostly used for generating the gas-phase ions from the eluting peptides of liquid phase sample, which are then further separated based on the ion's mass-to-charge ratio (m/z) by the mass analyzer. A detector then registers the m/z values of the ions present in the sample. Instrument control, sample acquisition and data analysis is managed by the software and associated electronics.

For a typical ion trap (LTQ-Linear Trap Quadrupole) or FTICR (Fourier Transform Ion Cyclotron Resonance) type of mass spectrometer, mass spectra are recorded in the positive ion mode following electrospray ionization at a capillary temperature of 275 °C, spray voltage of 4.5 kV, and sheath gas, auxiliary gas, and sweep gas set to 35, 10, and 15 arbitrary units, respectively. The scan range for recording mass-to-charge ratio (m/z) values (in MS mode) of peptides is between 400 and 2,000. The instrument is tuned automatically to adjust the voltages of capillary, tube lens, multipoles and other lenses with angiotensin I (5 pmol/µl) in 50 % mobile phase A and 50 % mobile phase B. Under standard tune conditions of 3 µl/min flow rate, the relative abundance of m/z 432.899 (+3 charge state) is 100 %, while that of m/z 648.86 (+2 charge state) is around 60 % with an injection time of <20 ms and ion intensity of >4 E5. If the targeted sample peptides are predicted to be that of lower charge states, then tuning can be adjusted to maximize the level of +2 charge state, i.e., m/z 648.86. This is accomplished by adjusting the capillary and tube lens voltages. The amino acid sequence of an isolated gas phase peptide ion is determined by collision induced dissociation (CID) in a two-dimensional mass spectrometry (MS/MS) analysis, also referred to as tandem mass spectrometry.

Collision-induced dissociation (CID) tandem mass spectrometry (normalized collision energy 35 %) is used in data-dependent mode to switch automatically between MS (which records m/z values for intact peptide ions), and four CID-MS/MS scans to obtain sequence information of the digestion products. Each precursor ion is isolated using the data-dependent acquisition mode with isolation width of m/z 3.0 to select automatically and sequentially a specific ion (starting with the most intense ion) from the first MS scan. Next CID-MS/MS scan was performed on second most abundant ion, remaining two CID-MS/MS scans were performed on the third and fourth most abundant ions, respectively, in order to obtain optimal sequence coverage. Acquiring MS scan in profile mode and MS/MS scans in centroid mode saves disc space without compromising data quality. The full mass spectra in the FTICR mode (m/z 400–2,000) is acquired at 100,000 mass resolution to determine the accurate mass and charge states of the precursor ions generated under similar conditions. In some cases, the tandem mass spectra of targeted peptides may not be acquired as they are masked by other co-eluting peptides with a more intense ion signal. Such peptides are sequenced using the targeted data-dependent scan method in which the MS/MS scans were performed for a set of precursor ions that matched with the masses of targeted peptides.

3.4 Data Analysis The MS instrument returns LTQ or LTQ-FT raw data files that contain the mass spectral data. These may be analyzed by Bioworks, that features the SEQUEST™ [20] protein search algorithm for protein/peptide identification. The predicted peptides in the search analysis are verified for the fragmentation pattern in the MS/MS spectra. Alternatively, the data can also be uploaded to the Mascot server [21], and analyzed for peptides and corresponding protein identification.

Data analysis and sequence interpretation of a typical tryptic peptide from DTT treated sample along with fragmentation data obtained by collision-induced dissociation (CID) is shown in Fig. 1. The interpretation of DTT-omitted or unreduced protein sample requires making assumptions and validation of these assumptions by high mass accuracy measurements, especially if the disulfide bond links two different tryptic peptides. At times, more than one MS-based dissociation pathway is employed to identify the given disulfide bond. For example, in case of multiple disulfide-containing peptides, high mass accuracy measurements generally provide the initial identification of S-S linkage. The sequence information of the associated peptides can be determined by comparison of peptide fragmentation patterns obtained by collision induced dissociation (CID) and electron transfer dissociation (ETD) pathways [22, 23].

In addition, reduction of disulfide linkage alters the peptide mobility on reverse phase C18 column, so comparison of peptide mobilities in the unreduced and reduced states can be an attractive possibility for disulfide bond detection.

4 Notes

1. 8 M urea, 0.4 M ammonium bicarbonate solution also works well for denaturation.

2. 1 % ProteaseMAX™ Surfactant solution can be stored 12–18 months at –20 °C in small aliquots. Avoid freezing and thawing cycles.

3. Precipitation helps in eliminating storage buffer and facilitates complete digestion of the protein sample. If the protein is not salt-free, it can also be desalted on Sephadex G-25 or G-50 microfuge columns.

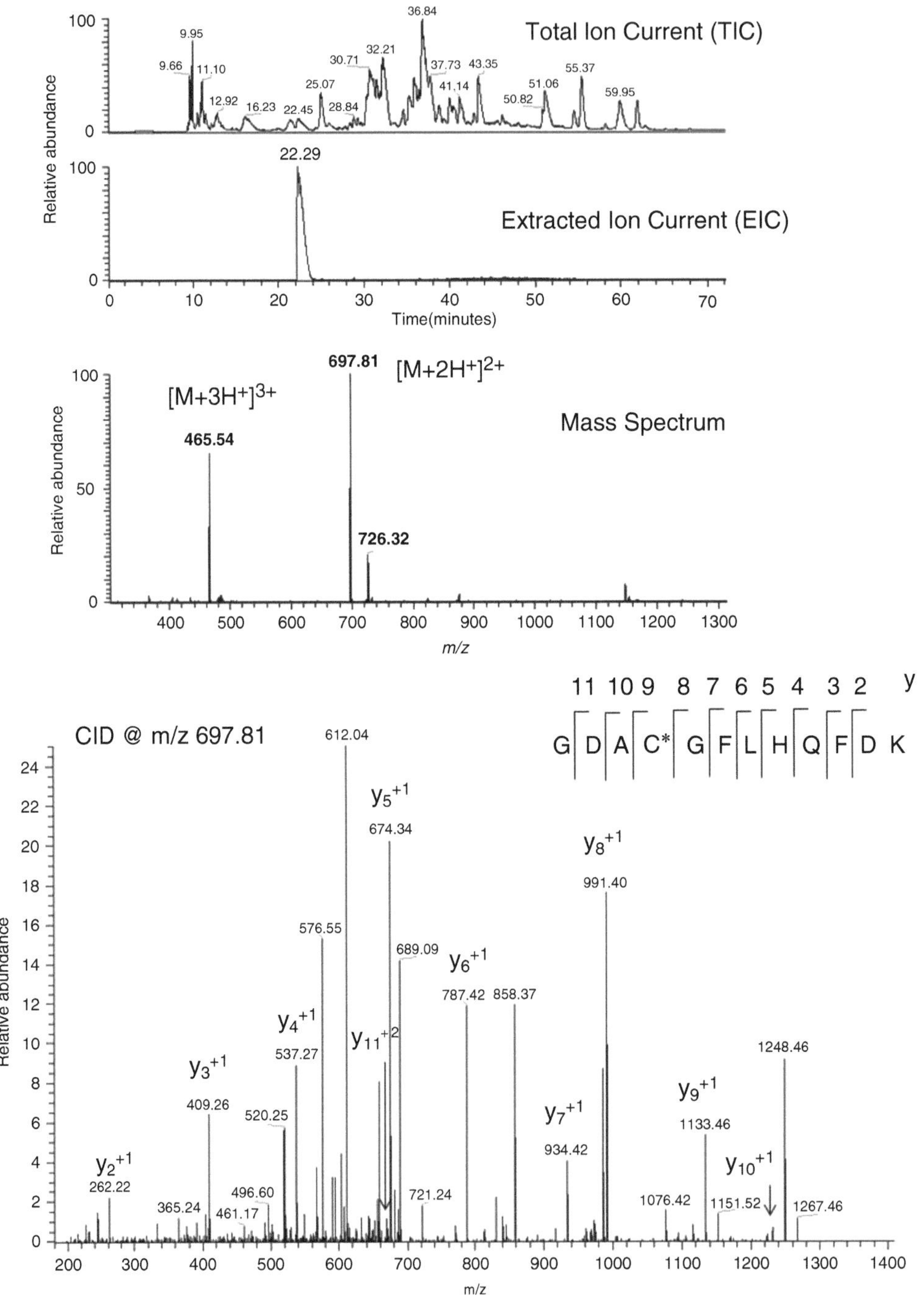

Fig. 1 LC-FTMS/MS analysis of tryptic peptide containing carbomidomethylated cysteine. (**a**) *Top panel* depicts the total ion current (TIC, all the ions eluted from reverse-phase column), *middle panel* shows the extracted ion current (EIC) for *m/z* 697.81 that corresponds to tryptic peptide GDAC*GFLHQFDK with a carbamidomethylated cysteine, and the *bottom panel* shows the mass spectrum of the extracted ion current. Note the presence of doubly and triply charged peptide ions in the spectrum. The carbamidomethylated status of cysteine is denoted by *asterisk*. (**b**) MS/MS spectrum of the tryptic peptide precursor ion with *m/z* 697.81. The fragment ions specific to peptide bond breakage, i.e., y_n ($n = 2$–10) (sharing common C-terminal end) depending on charge retention are represented on the spectrum. The complementary b_n ($n = 3$–10) ion series (sharing common N-terminal end) although observed in the spectrum, are not labeled to avoid cluttering. The sequence of peptide is deduced from the observed *m/z* values of fragment ions

4. A concentration of 1 μg/μl seems to be optimal for denaturation treatment.

5. The temperature and incubation time depends on the number of cysteines expected to be present in the protein and its compactness. The greater the number of cysteines and more compact the molecule, the higher the temperature and longer the incubation time.

6. Optimal binding of protein to ZipTip may also require a chaotropic agent (e.g., urea or guanidine-HCl at a final concentration of approximately 1–4 M). If sample does not already contain a chaotropic salt, add a few drops of salt before binding. In the case of excess detergent, dilute sample with 0.1 % TFA to achieve acceptable binding conditions, e.g., SDS (<0.1 %), Triton® (<1 %), and Tween® (<0.5 %).

7. Sample recovery can be improved (at the expense of concentration) by increasing elution volume to 10 μl or by performing multiple elutions. The volume of eluted sample can be reduced by drying in SpeedVac without adversely affecting the quality.

8. Reverse phase high performance liquid chromatography (RP-HPLC) is the approach of choice for primary separation of peptides to reduce complexity prior to MS analysis. In addition to the separation of peptides, the increasing percentage of organic component in the gradient elution also augments the ionization efficiency of eluting peptides in the MS.

References

1. Gilmartin GM (2005) Eukaryotic mRNA 3′ processing: a common means to different ends. Genes Dev 19(21):2517–2521. doi:10.1101/gad.1378105

2. Hunt AG, Xing D, Li QQ (2012) Plant polyadenylation factors: conservation and variety in the polyadenylation complex in plants. BMC Genomics 13:641. doi:10.1186/1471-2164-13-641

3. Hunt AG, Xu R, Addepalli B, Rao S, Forbes KP, Meeks LR, Xing D, Mo M, Zhao H, Bandyopadhyay A, Dampanaboina L, Marion A, Von Lanken C, Li QQ (2008) Arabidopsis mRNA polyadenylation machinery: comprehensive analysis of protein-protein interactions and gene expression profiling. BMC Genomics 9:220. doi:10.1186/1471-2164-9-220

4. Mandel CR, Bai Y, Tong L (2008) Protein factors in pre-mRNA 3′-end processing. Cell Mol Life Sci 65(7–8):1099–1122. doi:10.1007/s00018-007-7474-3

5. Zhang J, Addepalli B, Yun KY, Hunt AG, Xu R, Rao S, Li QQ, Falcone DL (2008) A polyadenylation factor subunit implicated in regulating oxidative signaling in Arabidopsis thaliana. PLoS One 3(6):e2410. doi:10.1371/journal.pone.0002410

6. Buchanan BB, Balmer Y (2005) Redox regulation: a broadening horizon. Annu Rev Plant Biol 56:187–220. doi:10.1146/annurev.arplant.56.032604.144246

7. Bykova NV, Rampitsch C (2013) Modulating protein function through reversible oxidation: redox-mediated processes in plants revealed through proteomics. Proteomics 13(3–4):579–596. doi:10.1002/pmic.201200270

8. Paulsen CE, Carroll KS (2010) Orchestrating redox signaling networks through regulatory cysteine switches. ACS Chem Biol 5(1):47–62. doi:10.1021/cb900258z

9. Winterbourn CC (2008) Reconciling the chemistry and biology of reactive oxygen species. Nat Chem Biol 4(5):278–286. doi:10.1038/nchembio.85

10. Murray CI, Van Eyk JE (2012) Chasing cysteine oxidative modifications: proteomic tools for

characterizing cysteine redox status. Circ Cardiovasc Genet 5(5):591. doi:10.1161/CIRCGENETICS.111.961425

11. Bachi A, Dalle-Donne I, Scaloni A (2013) Redox proteomics: chemical principles, methodological approaches and biological/biomedical promises. Chem Rev 113(1):596–698. doi:10.1021/cr300073p

12. Chouchani ET, James AM, Fearnley IM, Lilley KS, Murphy MP (2011) Proteomic approaches to the characterization of protein thiol modification. Curr Opin Chem Biol 15(1):120–128. doi:10.1016/j.cbpa.2010.11.003

13. Han B, Hare M, Wickramasekara S, Fang Y, Maier CS (2012) A comparative 'bottom up' proteomics strategy for the site-specific identification and quantification of protein modifications by electrophilic lipids. J Proteomics 75(18):5724–5733. doi:10.1016/j.jprot.2012.07.029

14. Izquierdo-Alvarez A, Martinez-Ruiz A (2011) Thiol redox proteomics seen with fluorescent eyes: the detection of cysteine oxidative modifications by fluorescence derivatization and 2-DE. J Proteomics 75(2):329–338. doi:10.1016/j.jprot.2011.09.013

15. Gallegos-Perez JL, Rangel-Ordonez L, Bowman SR, Ngowe CO, Watson JT (2005) Study of primary amines for nucleophilic cleavage of cyanylated cystinyl proteins in disulfide mass mapping methodology. Anal Biochem 346(2):311–319. doi:10.1016/j.ab.2005.08.003

16. Wefing S, Schnaible V, Hoffmann D (2006) SearchXLinks. A program for the identification of disulfide bonds in proteins from mass spectra. Anal Chem 78(4):1235–1241. doi:10.1021/ac051634x

17. Bach RD, Dmitrenko O, Thorpe C (2008) Mechanism of thiolate-disulfide interchange reactions in biochemistry. J Org Chem 73(1):12–21. doi:10.1021/jo702051f

18. Addepalli B, Hunt AG (2008) Ribonuclease activity is a common property of Arabidopsis CCCH-containing zinc-finger proteins. FEBS Lett 582(17):2577–2582. doi:10.1016/j.febslet.2008.06.029

19. Addepalli B, Limbach PA, Hunt AG (2010) A disulfide linkage in a CCCH zinc finger motif of an Arabidopsis CPSF30 ortholog. FEBS Lett 584(21):4408–4412. doi:10.1016/j.febslet.2010.09.043

20. Eng JK, McCormack AL, Yates Iii JR (1994) An approach to correlate tandem mass spectral data of peptides with amino acid sequences in a protein database. J Am Soc Mass Spectrom 5(11):976–989. doi:10.1016/1044-0305(94)80016-2

21. Cottrell JS (2011) Protein identification using MS/MS data. J Proteomics 74(10):1842–1851. doi:10.1016/j.jprot.2011.05.014

22. Mo J, Tymiak AA, Chen G (2013) Characterization of disulfide linkages in recombinant human granulocyte-colony stimulating factor. Rapid Commun Mass Spectrom 27(9):940–946. doi:10.1002/rcm.6530

23. Wu SL, Jiang H, Lu Q, Dai S, Hancock WS, Karger BL (2009) Mass spectrometric determination of disulfide linkages in recombinant therapeutic proteins using online LC-MS with electron-transfer dissociation. Anal Chem 81(1):112–122. doi:10.1021/ac801560k

Chapter 11

Transient Expression Using Agroinfiltration to Study Polyadenylation in Plants

Carol Von Lanken and Arthur G. Hunt

Abstract

An important facet of the study of polyadenylation in plants entails the use of transient expression to characterize poly(A) sites and the subunits of the polyadenylation complex. In this report, a simple and adaptable approach to transient expression in *Nicotiana benthamiana* leaves is described. This approach has been used to evaluate protein–protein interactions amongst different plant polyadenylation factor subunits and to study the usage and behavior of individual poly(A) sites from plant genes.

Key words Agroinfiltration, Transient expression, Polyadenylation sites, Protein–protein interactions

1　Introduction

The characterization of mRNA 3′ end formation in plants involves the study of several questions that are facilitated by the use of transient expression. The potential utility of transient expression approaches is considerable, and includes the qualitative and quantitative assessment of poly(A) sites and signals (e.g., [1]), the subcellular locations of individual polyadenylation-related proteins [2, 3], and the interactions between hypothetical and known polyadenylation-related proteins (as in Rao et al. [2]). These considerations make the availability of simple and versatile methods an important and going concern.

In the past several years, our group has adapted genes that encode fluorescent proteins for the study of several aspects of mRNA polyadenylation in plants. In particular, we have used "native" fluorescent protein reporters as well as proteins containing a nuclear localization signal to study the properties of poly(A) sites and signals as well as of proteins that are part of the polyadenylation complex. These various genes have been used in conjunction with *Agrobacterium*-mediated transient infection to rapidly assess different parameters. In this report, these methods are

Arthur G. Hunt and Qingshun Quinn Li (eds.), *Polyadenylation in Plants: Methods and Protocols*, Methods in Molecular Biology, vol. 1255, DOI 10.1007/978-1-4939-2175-1_11, © Springer Science+Business Media New York 2015

described, as are variations that make use of the different manifestations of fluorescent reporters that are available [4, 5]. Briefly, *Agrobacterium tumefaciens* cells carrying a suitable binary vector are grown, treated with acetosyringone, and infiltrated into leaves of young *Nicotiana benthamiana* plants. After a suitable incubation, plants are assessed using light microscopy, or RNA is isolated from infiltrated leaves for further analysis.

2   Materials

2.1 Agroinfiltration

Luria Broth liquid media: per liter, 10 g tryptone, 5 g yeast extract, 10 g NaCl, dissolved in deionized water and sterilized by autoclaving.

Luria Broth agar plates (1.5 % agar) with Kanamycin (50 mg/L) and Rifampicin (25 mg/L).

Rifampicin (25 mg/mL) in methanol.

Kanamycin (50 mg/mL) in Milli-Q water.

1 M $MgCl_2$.

1 M MES, pH 5.6.

MES infiltration buffer (10 mM $MgCl_2$, and 10 mM MES, pH 5.6).

100 mM acetosyringone in 70 % ethanol.

Spectrophotometer.

1 mL Syringes.

20 G Needle.

3-Week-old *Nicotiana benthamiana* plants (Fig. 1).

15 mL Sterile culture tubes.

Spatula.

Fig. 1 Photograph of a typical 3-week-old *N. benthamiana* plant used for transient expression studies

2.2 Microscopy	Zeiss Axioplan 2 HB100 microscope with differential contrast filters. Zeiss AxioCam MRc5 camera attached to the microscope. AxioVision software. Microscope slides and cover glass. Agroinfiltrated leaves. Water.
2.3 RNA Isolation	Agroinfiltrated leaves Trizol. Mortar and pestle. Liquid Nitrogen. Chloroform. Isopropyl alcohol. RNAse-free water.

3   Methods

3.1 Growth of Agrobacterium tumefaciens (See Note 1) *3.1.1 Growing Agrobacterium on Plates (See **Note 2**)*	1. Grow suitably transformed *Agrobacterium tumefaciens* LBA 4404 cells in liquid Luria Broth containing rifampicin and kanamycin (*see* **Note 1**). Spread 100–200 µL of a stationary phase culture on Luria Broth agar plates containing rifampicin and kanamycin. Incubate at 28 °C for 2–3 days. 2. When the plates are covered with *Agrobacterium*, scrape the bacteria from the plates with a sterile spatula and put in 15 mL culture tubes. Resuspend in MES buffer (10 mm MgCl₂, 10 mm MES, pH 5.6) and adjust the OD (600 nm) to 0.6 (*see* **Note 3**).
*3.1.2 Growing Agrobacterium in Liquid Cultures (See **Note 2**)*	1. Beginning with the starter culture used in Subheading 3.1.1, **step 1**, inoculate 25 mL of LB-rif-kan with 25–100 µL of starter culture and incubate at 28 °C for 2–3 days. 2. Pellet the cells in 10 mL of culture in a clinical centrifuge for 5 min on a setting of 7. Remove the supernatant and resuspend the bacterial pellet in 2 mL of MES buffer. Adjust the OD₆₀₀ to 0.6 using MES buffer as needed (*see* **Note 3**).
3.2 Agroinfiltration	1. Add acetosyringone to the diluted *Agrobacterium* cultures to a final concentration of 150 µM. Incubate at room temperature for 2–3 h. 2. If the experimental design includes co-infiltration of two or three constructs in one plant, combine the acetosyringone-treated cultures at the end of the 3 h incubation in acetosyringone. The ratio for mixing the cultures for coinfiltration is 1:1

Fig. 2 (**a**) Photograph showing the nicking of the leaf. (**b**) Photograph showing the appearance of a leaf partway through the infiltration process. The contrast between the saturated and unsaturated parts of the leaf is readily apparent

for a co-infiltration of two constructs, and 1:1:1 one involving three constructs.

3. After the incubation in acetosyringone, infiltrate the plants as follows. Select three consecutive, fully expanded leaves on each plant (*see* **Note 4**). Using the needle, make a small hole on the abaxial surface of each leaf that is neither on nor directly adjacent to a large vein in the leaf (Fig. 2a). Using a 1 cm^3 syringe filled with the appropriate culture or mixture of cultures, infiltrate each leaf by placing the syringe over the nick and slowly injecting the culture into the abaxial surface of the leaf while supporting the leaf with your hand (Fig. 2b; *see* **Note 5**).

4. When all of the leaves are infiltrated, the plants are returned to their pre-infiltration growing conditions.

3.3 Microscopy

1. After 3 and 7 days, remove small leaf sections and place in a drop of water between a slide and coverslip (Fig. 3). Examine samples under UV light in the epifluorescence Zeiss Axioplan 2 HB100 microscope with additional DIC filter (differential contrast) that can visualize GFP (excites at 488 nm) and RFP (excites at 543 nm). Capture the images with a Zeiss camera with AxioCam MRC5 using AxioVision software.

2. Visualize GFP expression with the green filter (i.e., D470/40; D535/40; beam splitter 500 DLCP) and RFP expression with the red filter (i.e., HQ545/30X; em: HQ610/75M, Q570LPBS) (*see* **Note 6**).

3.4 RNA Isolation

1. Weigh out 1 g of leaf tissue from agroinfiltrated leaves. Freeze with liquid nitrogen, and grind to a fine powder. Add 10 mL of Trizol to the frozen tissue, mixing well until liquid the powder is thawed (*see* **Note 7**).

Fig. 3 Photograph of the arrangement of leaf explant on the cover slip

2. Allow the thawed Trizol/leaf tissue to incubate at room temperature for 5 min. Add 0.2 mL chloroform per mL of Trizol used in **step 1**. Shake the tube vigorously for 15 s, followed by incubation at room temperature for 3 min.

3. Centrifuge at 12,000 × g for 15 min.

4. Carefully transfer the upper aqueous layer to a fresh tube. Add 0.5 mL isopropyl alcohol per mL of transferred aqueous layer. Store at –20 °C overnight.

5. Centrifuge at 12,000 × g for 10 min. Remove supernatant. Wash the pellet with 1 mL of 75 % ETOH. Vortex and centrifuge at 7,500 × g for 5 min.

6. Discard the supernatant and air-dry pellets for 5–10 min.

7. Dissolve RNA in 50–200 μL of RNase-free water. Incubate at 55–60 °C for 10 min (*see* **Note 8**). This is the RNA that may be used for further processing and analysis.

4   Notes

1. We typically use *Agrobacterium tumefaciens* strain LBA 4404 [6]. However, this method should be equally useful with most other *Agrobacterium* strains. The antibiotics used will be determined by the *Agrobacterium* strain (LBA4404 carries a rifampicin resistance determinant) and the marker(s) resident on the binary vector used for plant gene transfer. Kanamycin resistance is used here because the standard laboratory vectors in use [4, 5] a kanamycin resistance determinant for use in bacterial cultures.

2. Growing the *Agrobacteria* on plates provides a greater quantity of cells. However, the liquid cultures were much easier to work with than scraping the solid media plates. Ten milliliters of liquid culture or less is sufficient to concentrate by centrifugation and make appropriate dilutions.

3. The amount of culture required for incubation is 3–4 mL per leaf × 3 leaves/plant × the number of plants replicates. Assume 12 mL of culture per plants for a single construct. For two constructs, 6 mL of each culture would be mixed just prior to infiltration for each plant, and for a triple construct 4 mL of each culture would be mixed per plant just prior to infiltration.

4. As plants mature, the leaf epidermis seems tougher, and the infiltrations become increasingly difficult to get complete saturation of the leaf.

5. The entire leaf should have a wet, saturated appearance (as in Fig. 2b), before moving on to the next leaf. To get complete saturation of a leaf, it is sometimes needful to repeat the process multiple times until the culture could no longer be infiltrated into the leaf.

6. These parameters are specific for the instruments (Zeiss Axioplan 2 HB100 and Zeiss AxioCam MRc5) used by our laboratory. Different parameters will be required for other instruments. Of course, it is possible to utilize more elaborate microscopic tools (such as confocal microscopy) as the experiment demands.

7. It may take a few minutes for the slurry of frozen tissue to liquefy in the presence of Trizol. None of the tissue should be allowed to thaw without being incorporated into the Trizol solution.

8. To completely dissolve the RNA, it may be necessary to pipette RNA solutions following the 10-min heating at 55–60 °C.

Acknowledgements

This work was supported by the US National Science Foundation (MCB-0313472 and IOS-0817818).

References

1. Zhang J, Addepalli B, Yun KY, Hunt AG, Xu R, Rao S, Li QQ, Falcone DL (2008) A polyadenylation factor subunit implicated in regulating oxidative signaling in Arabidopsis thaliana. PLoS One 3(6):e2410

2. Rao S, Dinkins RD, Hunt AG (2009) Distinctive interactions of the Arabidopsis homolog of the 30 kD subunit of the cleavage and polyadenylation specificity factor (AtCPSF30) with other polyadenylation factor

subunits. BMC Cell Biol 10:51. doi:10.1186/1471-2121-10-51

3. Meeks LR, Addepalli B, Hunt AG (2009) Characterization of genes encoding poly(A) polymerases in plants: evidence for duplication and functional specialization. PLoS One 4(11):e8082. doi:10.1371/journal.pone.0008082

4. Chakrabarty R, Banerjee R, Chung SM, Farman M, Citovsky V, Hogenhout SA, Tzfira T, Goodin M (2007) PSITE vectors for stable integration or transient expression of autofluorescent protein fusions in plants: probing Nicotiana benthamiana-virus interactions. MPMI 20(7):740–750. doi:10.1094/MPMI-20-7-0740

5. Goodin MM, Dietzgen RG, Schichnes D, Ruzin S, Jackson AO (2002) pGD vectors: versatile tools for the expression of green and red fluorescent protein fusions in agroinfiltrated plant leaves. Plant J 31(3):375–383

6. Hoekema A, Hirsch PR, Hooykaas PJJ, Schilperoort RA (1983) A binary plant vector strategy based on separation of vir- and T-region of the Agrobacterium tumefaciens Ti-plasmid. Nature 303(5913):179–180

Chapter 12

A 3′ RACE Protocol to Confirm Polyadenylation Sites

Liuyin Ma and Arthur G. Hunt

Abstract

A routine procedure in the study of polyadenylation involves the determinations of the junctions of the mRNA body and the poly(A) tail (the poly(A) site). This is typically accomplished by selectively amplifying the 3′-ends of cDNAs (3′-Rapid Amplification of cDNA Ends, or 3′-RACE), followed by sequencing of individual clones. We have developed a modification of the standard 3′-RACE protocol that couples high specificity with the ability to characterize thousands (or more) of individual sequences. This protocol may be used for numerous purposes, including the confirmation of results obtained using high throughput sequencing technologies.

Key words 3′ RACE protocol, Polyadenylation sites, Ion torrent, MiSeq

1 Introduction

A routine and ongoing concern in the study of polyadenylation in plants involves the determination of the locations along a pre mRNA at which the RNA is processed and polyadenylated. This has been true for studies as diverse as experimental determinations of polyadenylation signals [1–6] to genome-wide characterizations of poly(A) site choice using high throughput DNA sequencing technologies [7–9]. While the specific experimental approaches have evolved over the years as cloning and DNA amplification technologies have grown more sophisticated, the standard method of choice remains some variation of the so-called Rapid Amplification of cDNA Ends (RACE) technique.

RACE, also described as "one-sided" PCR or "anchored PCR," is a powerful technique using messenger RNA as template to amplify the nucleic acid sequences at either the 3′ (3′-RACE) or the 5′ (5′-RACE) end of the mRNA [10]. For studies of mRNA polyadenylation, 3′-RACE is the method of interest. In 3′-RACE, cDNA copies of a population of mRNAs are produced using oligo-dT as a primer for reverse transcriptase, and sequences that correspond to the 3′-ends (including the poly(A) sites) of specific

Arthur G. Hunt and Qingshun Quinn Li (eds.), *Polyadenylation in Plants: Methods and Protocols*, Methods in Molecular Biology, vol. 1255, DOI 10.1007/978-1-4939-2175-1_12, © Springer Science+Business Media New York 2015

mRNAs are amplified using some combination of gene-specific and library-specific primers. Gene-specific primers typically are derived from sequences upstream from the poly(A) site, while library-specific primers are homologous to the primers used for cDNA synthesis. Numerous variations of this basic theme are possible; these include the use of nested primers and multiple rounds of amplification, initial attachment of RNA adapters to the population of mRNAs to provide a non-oligo-dT platform at which cDNA synthesis may be primed, and all of the scope of PCR amplification strategies that abound in the literature [11–14].

This report describes a 3′-RACE protocol that has been used to confirm poly(A) sites identified in several genome-wide characterizations of polyadenylation in Arabidopsis. The protocol (illustrated in Fig. 1) incorporates a number of novel strategies and steps, and at its end results in the return of thousands (or more) of sequences specific for the gene of interest. Thus, beginning with purified RNA, cDNA is synthesized using a biotinylated and anchored oligo-dT primer into which is built a sequence designed to be compatible with a high throughput sequencing platform. The cDNA is then copied using an adaptation of the so-called asymmetric amplification method [15], so as to enrich for cDNAs derived just from the gene of interest. The original cDNA is then removed using the biotin moiety at the 5′ end of the reverse transcription primer, and the asymmetric amplification products amplified using standard PCR. High throughput sequencing adapters are then added to the PCR products using a low-cycle round of amplification, and the products purified and submitted for sequencing. The end result is a collection of thousands of sequences specific for the gene of interest.

2 Materials

1. Nuclease-free water (not DEPC-Treated)

2. Thin-walled microcentrifuge tubes (0.2 mL or 0.5 mL)

3. Primers for reverse transcription and PCR amplification

4. 5× RevertAid reverse transcriptase first strand buffer

5. RevertAid Reverse Transcriptase

6. 10× dNTP mix for reverse transcriptase: 10 mM each of dATP, dTTP, dGTP, and dCTP, all in water

7. 100 mM DTT (dithiothreitol)

8. RNase inhibitor, Human Placenta (1 U/µL)

9. Programmable thermocycler with heated lid

10. RNase H, 5 U/µL

11. RNase A/T1 Mix—2 mg/mL RNAse A, 5,000 U/mL RNAse T1 in 50 mM Tris–HCl, pH 7.4 + 50 % (v/v) glycerol

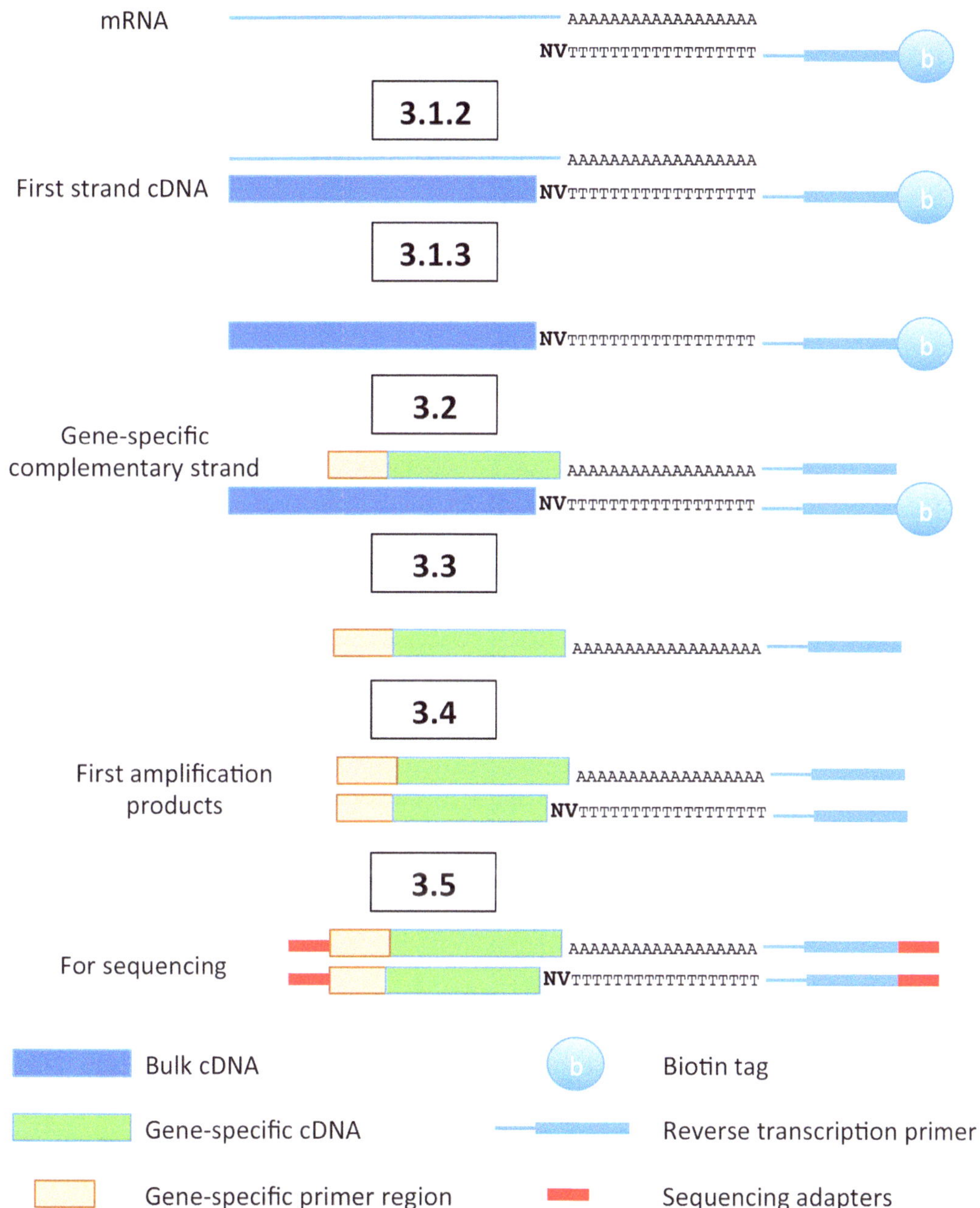

Fig. 1 Illustration of the workflow for the production of 3'-RACE high throughput sequencing samples. Each nucleic acid moiety (mRNA, cDNA, etc.) is labeled at the *left* at its first occurrence in the workflow. The sequence details for the 3' end of reverse transcriptase primer that is responsible for hybridizing to the poly(A) tract of the mRNA, and for anchoring the primer to the mRNA-poly(A) junction, is shown as an extension of the reverse transcription primer. Protocol steps associated with the illustrated workflow are indicated in the respective *text boxes*. "Bulk cDNA" indicates the complete collection of cDNAs synthesized at the outset. "Gene-specific cDNA" indicates the results of asymmetric amplification using a gene-specific primer. As illustrated, the transition or conversion of bulk to gene-specific products is accomplished by the manipulations described in Subheadings 3.2 and 3.3

12. PCR Purification Kit—includes spin columns and buffers for column purification, washing, and elution of PCR products

13. Phire® Hot Start II DNA Polymerase (1 U/μL) and 5× buffer provided with enzyme

14. 10× dNTP mix for PCR: 2.5 mM each of dATP, dTTP, dGTP, and dCTP, all in water

15. Streptavidin magnetic beads, 4 mg/mL

16. Magnetic separation rack

17. Agarose, molecular biology grade

18. TBE: 45 mM Tris-borate, pH 8.0, 1 mM Na-EDTA

19. Apparatus for agarose gel electrophoresis (gel box and casting tray, power supply)

20. Gel Extraction Kit for DNA purification (includes spin columns and buffers for solubilization of agarose gel fragments, column purification, washing, and elution of PCR products)

21. Sterile scalpel for dissection of agarose gels

3   Methods

3.1   cDNA Synthesis

1. Mix 14 μL of purified RNA (*see* **Note 1**) with 1 μL of the biotinylated reverse transcription primer (100 pmol/μL; *see* **Note 2**) heat to 65 °C and incubate for 5 min in a heat block, chill on ice for 2 min.

2. Add 5 μL 5× RevertAid reverse transcriptase first strand buffer (*see* **Note 3**), 2.5 μL 10× dNTPs for reverse transcription, 1 μL 100 mM DTT, and 0.5 μL RNase Inhibitor to the tube from **step 1**. Bring the tube to 42 °C in a thermocycler with a heated lid, then add 1 μL RevertAid reverse transcriptase and incubate for 2 h at 42 °C.

3. Heat the mixture to 70 °C for 5 min. Add 1 μL RNase H and RNase A/T1 (*see* **Note 4**). Incubate at 37 °C for 1 h.

4. Purify the cDNA using the PCR Purification Kit. Elute into 50 μL of the elution buffer provided with the PCR purification kit.

3.2   Asymmetric PCR Amplification

1. In a thin-wall microcentrifuge tube (*see* **Note 5**), mix the following: 5 μL of 5× Phire Hot Start II Polymerase buffer (*see* **Note 6**), 2.5 μL of 10× dNTPs for PCR, 1 μL of gene specific forward primer (10 pmol/μL; *see* **Note 7**), 1 μL of cDNA from Subheading 3.1, **step 4**, and water to bring the volume to 24.5 μL.

2. Add 0.5 μL of Phire Hot Start II DNA Polymerase and mix the contents of the tube by pipetting up and down 6–10 times.

3. Perform PCR for 20–50 cycles (*see* **Note 8**) using the following cycle parameters: initial denaturation for 30 s at 98 °C, cycles consisting of denaturation for 15 s at 98 °C, annealing for 15 s at 60 °C (*see* **Note 9**), extension for 2 min (*see* **Note 9**) at 72 °C, and a final extension for 10 min at 72 °C.

4. Purify the amplified DNA using the QIAquick PCR Purification Kit. Elute into 50 µL of the supplier-provided elution buffer.

3.3 Streptavidin Bead Clean Up (See Note 10)

1. Place 10 µL of streptavidin magnetic beads into a 500 µL thin-wall microcentrifuge tube. Add 100 µL of sterile water, mix, capture beads with magnet, and discard the supernatant. Repeat the water wash step and discard the supernatant.

2. Add the entire asymmetric PCR product sample (Subheading 3.2, **step 4**) to the washed beads, mix, and incubate at room temperature for 10 min. Separate the beads using the magnetic separation rack, remove and save supernatant.

3.4 PCR Amplification

1. In a thin-wall microcentrifuge tube (*see* **Note 5**), mix the following: 5 µL of 5× Phire Hot Start II Polymerase buffer (*see* **Note 6**), 2.5 µL of 10× dNTPs for PCR, 1 µL of a gene specific forward primer (10 pmol/µL; *see* **Note 11**), 1 µL of the library-specific reverse primer (10 pmol/µL; *see* **Note 11**), 1 µL of cDNA from Subheading 3.3, **step 2**, and water to bring the volume to 24.5 µL.

2. Add 0.5 µL of Phire Hot Start II DNA Polymerase and mix the contents of the tube by pipetting up and down 6–10 times.

3. Perform PCR for 20–30 cycles (*see* **Note 12**) using the following cycle parameters: initial denaturation for 30 s at 98 °C, cycles consisting of denaturation for 15 s at 98 °C, annealing for 15 s at 60 °C (*see* **Note 9**), extension for 2 min (*see* **Note 9**) at 72 °C, and a final extension for 10 min at 72 °C.

4. Separate the PCR products on an agarose gel (0.8–1.5 %, cast and run in 1× TBE) (*see* **Note 13**). Select and excise the desired amplification products (*see* **Note 14**) and purify the DNA using the Qiagen Gel Extraction kit.

3.5 Addition of the Sequencing Adapter (See Note 15) and DNA Sequencing

1. In a thin-wall microcentrifuge tube (*see* **Note 5**), mix the following: 5 µL of 5× Phire Hot Start II Polymerase buffer (*see* **Note 5**), 2.5 µL of 10× dNTPs for PCR, 1 µL of a hybrid sequencing adapter/gene specific forward primer (10 pmol/µL; *see* **Note 16**), 1 µL of a hybrid sequencing adapter/library-specific reverse primer (*see* **Note 16**), 1 µL of PCR product from Subheading 3.4, **step 4**, and water to bring the volume to 24.5 µL.

2. Add 0.5 µL of Phire Hot Start II DNA Polymerase and mix the contents of the tube by pipetting up and down 6–10 times.

3. Perform PCR for no more than 10 cycles (*see* **Note 17**) using the following cycle parameters: initial denaturation for 30 s at 98 °C, cycles consisting of denaturation for 15 s at 98 °C, annealing for 15 s at 60 °C (*see* **Note 9**), extension for 2 min (*see* **Note 9**) at 72 °C, and a final extension for 10 min at 72 °C.

4. Purify the amplified DNA using the QIAquick PCR Purification Kit. Elute into 50 μL of the supplier-provided elution buffer.

5. Assess the PCR products on an agarose gel (0.8–1.5 %, cast and run in 1× TBE).

6. Submit the sample to a sequencing facility (*see* **Note 18**).

4   Notes

1. Typically, total RNA that has been purified using organic extraction and/or column purification (e.g., using systems analogous to the Qiagen RNA Purification kits) is used. RNA quantities will be dictated by the abundance of transcripts in the samples; this procedure has been used with RNA quantities ranging from about 1 μg to more than 30 μg of total RNA. Of course, poly(A)-enriched RNA may be used in place of total RNA; in such instances, quantities should be adjusted accordingly.

2. The reverse transcription primer has the general structure: 5′-biotin-$(X)_n$TTTTTTTTTTTTTTTTTTTVN-3′ (*see* Fig. 2). $(X)_n$ denotes a user-defined sequence that is specific for a particular experiment or project and may range from between 20 to as many as 60 nts in length. This sequence is typically used for the placement of the library-specific primer in subsequent PCR amplifications, and may be large enough to accommodate a set of overlapping primers for nested PCR. Regardless of these considerations, this sequence should be absent from the genome of the organism being studied and the respective library-specific primer binding sites should be chosen to optimize performance in PCR amplifications. Other considerations include the possibility that this sequence will constitute the first section of the high throughput sequencing reads (and thus will constrain the extent of sequence used to identify each clone; *see* **Notes 14** and **15**). Thus, one option is to place a part of a sequencing adapter for Illumina, Ion Torrent, or 454 sequencing at this position. At the 3′ end of the primer, the last two positions (VN) are intended to anchor the primer at the poly(A)–mRNA junction.

3. We use the RevertAid enzyme and associated buffers routinely. However, other sources of the enzyme may be substituted for the reverse transcription reactions; in such cases, the buffers and protocols provided by the supplier should be followed.

X_n – user-designated sequence (20-40 nts)

VN – anchoring sequence (promotes hybridization to the mRNA-poly(A) junction)

NN – random sequence (important for sequencing and analysis)

yyy – user-designated barcode (used to permit multiplexing of samples in a single sequencing run)

Biotin tag at 5′ end of RT primer

Fig. 2 Illustration of the primer design for the 5′-ends of the cDNA library (corresponding to the 3′ ends of the respective cDNAs), emphasizing the continuity represented by the user-designated sequence appended to the oligo-dT tract in the RT primer. Components of the primers are as indicated in the legend beneath the depictions of the primers

4. This combination of ribonucleases exceeds the inhibitory capacity of the RNAse inhibitor added to the reverse transcription reactions by more than tenfold. This treatment eliminates RNA (free and hybridized to the cDNA), allowing unfettered access of the cDNA to DNA primers and reducing the possibility of amplification of RNA by Taq polymerases, that may possess low levels of reverse transcriptase activity under some conditions [16, 17].

5. Typically, thin-walled 200 µL tubes are used for PCR amplifications that are performed in programmable thermocyclers. However, these may be replaced with 96 well microtiter plates with 200 µL wells if the sample number is large, and if the thermocycler can accommodate plates. Alternatively, larger (500 µL) tubes may be used if the thermocycler cannot accommodate 200 µL tubes.

6. Our laboratories routinely use the Phire II Taq DNA Polymerase for PCR amplifications. However, we have used other enzymes, hot-start and not, on occasion. Thus, the choice of enzyme (and associated buffers) is largely a matter of the preferences of users.

7. The gene-specific forward primer is designed to provide optimal performance in the amplification process. Of course, the thermodynamic properties of the predicted primer-template hybrids will factor into the PCR parameters. The genomic position of the gene-specific primer is usually chosen to yield an amplification product of between 100 and 1,000 bp.

8. The cycle number should be empirically determined, and the lowest number that yields the appropriate product chosen for further steps.

9. Annealing temperatures will be determined by the thermodynamic properties of the primer-template hybrid, and can usually be predicted using standard PCR primer design tools. The extension time is chosen to permit full extension of expected and possible (if unanticipated) products; this time will reflect both target length and the kinetic properties of the Taq DNA polymerase being used.

10. The cleanup step with streptavidin beads serves to remove the first strand cDNA. What remains is a population of single-stranded molecules that is highly enriched for target of interest. Removal of other, unrelated cDNAs eliminates the imbalance that exists due to the fact that all molecules can anneal to one primer (the library-specific one), but only a tiny fraction anneals to the gene-specific primer. Reduction of this imbalance tremendously improves subsequent PCR performance.

11. Primers are typically designed to be nested. However, the same primers as those used in Subheading 3.1, **step 1** (absent the biotin moiety) and Subheading 3.2, **step 1** may be used if desired.

12. The cycle number should be empirically determined, and the lowest number that yields the appropriate product chosen for further steps

13. The agarose concentrations and running conditions will be determined by the equipment available as well as the nature (e.g., sizes) of the expected and observed PCR products. Regardless, electrophoretic separation and documentation can be performed using any of a large number of apparatuses and formats.

14. PCR products may be purified individually, or as a group. The latter approach is suited for instances where multiple mRNA species are represented by 3′-RACE products that differ in size by less than 200 bp or so. If the latter alternative is chosen, different individual products can be "resolved" computationally during the analysis of sequences returned from sequencing centers (*see* **Note 18**).

15. An economical alternative to the cloning of DNA fragments obtained in Subheading 3.4, **step 4** and subsequent sequencing of individual clones is to sequence the PCR products directly

and en masse. For this, the gel-purified DNA fragments need to be adapted for high throughput sequencing. However, the PCR products from Subheading 3.4, **step 4** will not be suitable for high throughput DNA sequencing. Thus, appropriate sequencing adapters must be added. This is done with a brief PCR amplification following the methods described in Subheading 3.5.

16. The design of the hybrid sequencing/gene-specific and sequencing/library-specific primers is shown in Fig. 2. The extents of commonalities between these primers and the respective gene-specific and library-specific primers from Subheading 3.2, **step 1** and Subheading 3.4, **step 1** are such that the re-amplification can be conducted at similar annealing temperatures as for Subheading 3.4, **step 2**. As with most aspects of this protocol, there is room for flexibility and creativity on the parts of users in the design of these primers. Regardless, the end result should be PCR products that possess the necessary sequencing adapters.

17. The cycle number at this step should be empirically determined, with the fewest number of cycles needed to generate visible bands after gel electrophoresis being all that is needed. The goal of this step is to add the platform-specific sequencing adapters to the ends of the PCR products generated in Subheading 3.4.

18. DNA purified using the Qiagen kit (or similar systems) is suitable for sequencing on most platforms, and is usually submitted with no further workup. Details of the sequencing and analysis will rely heavily on the choice of platform, facility, and analysis software. However, sequencing returns on the orders of 1–10 million reads, of lengths between 100 and 250 nts, are entirely sufficient for successful completion of a given project. This extent of return is compatible with less-expensive technologies (MiSeq, Ion Torrent), and the sequence files are readily analyzed using lower-cost commercial software packages (such as Geneious [Biomatters LTD]) as well as web-based services (such as iPlant [http://www.iplantcollaborative.org/] and Galaxy [http://galaxyproject.org/]).

Acknowledgements

This work was supported by US National Science Foundation (IOS-081781). In addition, Liuyin Ma was a recipient of a scholarship from the China Scholarship Council.

References

1. Mogen BD, MacDonald MH, Graybosch R, Hunt AG (1990) Upstream sequences other than AAUAAA are required for efficient messenger RNA 3′-end formation in plants. Plant Cell 2:1261–1272

2. Mogen BD, MacDonald MH, Leggewie G, Hunt AG (1992) Several distinct types of sequence elements are required for efficient mRNA 3′ end formation in a pea rbcS gene. Mol Cell Biol 12(12):5406–5414

3. Rothnie HM, Reid J, Hohn T (1994) The contribution of AAUAAA and the upstream element UUUGUA to the efficiency of mRNA 3′-end formation in plants. EMBO J 13(9):2200–2210

4. Sanfacon H, Brodmann P, Hohn T (1991) A dissection of the cauliflower mosaic virus polyadenylation signal. Genes Dev 5(1):141–149

5. Wu L, Ueda T, Messing J (1993) 3′-end processing of the maize 27 kDa zein mRNA. Plant J 4(3):535–544

6. Wu L, Ueda T, Messing J (1994) Sequence and spatial requirements for the tissue- and species-independent 3′-end processing mechanism of plant mRNA. Mol Cell Biol 14(10):6829–6838

7. Sherstnev A, Duc C, Cole C, Zacharaki V, Hornyik C, Ozsolak F, Milos PM, Barton GJ, Simpson GG (2012) Direct sequencing of Arabidopsis thaliana RNA reveals patterns of cleavage and polyadenylation. Nat Struct Mol Biol 19(8):845–852. doi:10.1038/nsmb.2345

8. Thomas PE, Wu X, Liu M, Gaffney B, Ji G, Li QQ, Hunt AG (2012) Genome-wide control of polyadenylation site choice by CPSF30 in Arabidopsis. Plant Cell 24(11):4376–4388. doi:10.1105/tpc.112.096107

9. Wu X, Liu M, Downie B, Liang C, Ji G, Li QQ, Hunt AG (2011) Genome-wide landscape of polyadenylation in Arabidopsis provides evidence for extensive alternative polyadenylation. Proc Natl Acad Sci U S A 108(30):12533–12538. doi:10.1073/pnas.1019732108

10. Frohman MA, Dush MK, Martin GR (1988) Rapid production of full-length cDNAs from rare transcripts: amplification using a single gene-specific oligonucleotide primer. Proc Natl Acad Sci U S A 85(23):8998–9002

11. Frohman MA (1994) On beyond classic RACE (rapid amplification of cDNA ends). PCR Methods Appl 4(1):S40–S58

12. Frohman MA (2006) 3′-end cDNA amplification using classic RACE. CSH Protoc 2006(1): pii: pdb.prot4130. doi:10.1101/pdb.prot4130

13. Mizobuchi M, Frohman LA (1993) Rapid amplification of genomic DNA ends. Biotechniques 15(2):214–216

14. Zhang Y, Frohman MA (1998) Cloning cDNA ends using RACE. Methods Mol Med 13:81–105. doi:10.1385/0-89603-485-2:81

15. Shyamala V, Ames GF (1989) Amplification of bacterial genomic DNA by the polymerase chain reaction and direct sequencing after asymmetric amplification: application to the study of periplasmic permeases. J Bacteriol 171(3):1602–1608

16. Jones MD, Foulkes NS (1989) Reverse transcription of mRNA by Thermus aquaticus DNA polymerase. Nucleic Acids Res 17(20):8387–8388

17. Myers TW, Gelfand DH (1991) Reverse transcription and DNA amplification by a Thermus thermophilus DNA polymerase. Biochemistry 30(31):7661–7666

Genome-Scale Study of Polyadenylation in Plants

Chapter 13

Phage Display Library Screening for Identification of Interacting Protein Partners

Balasubrahmanyam Addepalli, Suryadevara Rao, and Arthur G. Hunt

Abstract

Phage display is a versatile high-throughput screening method employed to understand and improve the chemical biology, be it production of human monoclonal antibodies or identification of interacting protein partners. A majority of cell proteins operate in a concerted fashion either by stable or transient interactions. Such interactions can be mediated by recognition of small amino acid sequence motifs on the protein surface. Phage display can play a crucial role in identification of such motifs. This report describes the use of phage display for the identification of high affinity sequence motifs that could be responsible for interactions with a target (bait) protein.

Key words Phage display, Protein–protein interactions, High-throughput screening, Random combinatorial approach

1 Introduction

The primary transcripts of eukaryotic messenger RNAs undergo extensive processing events before their transport into the cytoplasm. One such event is addition of a poly(A) tail, widely referred to as polyadenylation or 3′ end processing. This event involves complex interplay of cis-elements and trans-acting protein factors [1–4]. Although the protein factors that perform polyadenylation in vitro are identified, this process is shown to be coupled and integrated with other steps of mRNA biogenesis in vivo including transcription, splicing and transcript termination [5, 6] through protein–protein interactions. Documentation of such interactions will not only help in understanding the polyadenylation machinery in a larger context of gene expression, it also can help in understanding the regulation of processing events mediated by developmental or cell signaling stimuli [6].

Phage display is a combinatorial technique employed to identify high affinity peptide motifs that could mediate protein–protein interactions. This technique entails expression and presentation of

Arthur G. Hunt and Qingshun Quinn Li (eds.), *Polyadenylation in Plants: Methods and Protocols*, Methods in Molecular Biology, vol. 1255, DOI 10.1007/978-1-4939-2175-1_13, © Springer Science+Business Media New York 2015

147

all/majority of the possible amino acid combinations of a given sequence (of a specified length) as a fusion with a bacteriophage coat protein, generally a filamentous phage such as M13. Ideally, one of the 20 amino acids substitutes each amino acid at a given position of the specified sequence in a random combinatorial manner to generate a phage displayed library of peptides. Screening such library will result in selection of a few peptide motifs that exhibit high affinity towards the surface (epitopes) of a bait protein. Phage display creates a physical linkage between the phage-displayed peptide sequences (of a given library) to the DNA that encodes amino acid sequence in a peptide. Thus, sequencing the DNA corresponding to the region of phage coat protein fusion will tell us the sequence of the affinity selected peptide.

This process of affinity selection of phage displayed peptides is termed as *panning*. Panning involves incubation of phage-displayed peptide library with an immobilized target protein (usually coated on plate), washing away the unbound phage, and eluting the specifically bound phage. The eluted phage is amplified and taken through repeated rounds of panning/amplification cycles to enrich the pool in favor of peptides that bind with high specificity. The bound peptides are characterized by DNA sequencing either after each round or after multiple cycles of selection. Smith [7] using filamentous phage as an expression vector initially introduced this concept. Since then, several variations of the method and associated applications (including exploration of protein–protein interactions) of phage display library have been published [8–18]. Phage display is considered as one of the versatile tools employed in identification of novel protein–protein interactions [19]. With phage display, we have been able to identify protein–protein contact points between some of *Arabidopsis* polyadenylation factors [20].

The filamentous phage M13 is the most commonly used vector because among other things, the phage can tolerate genome insertions in the nonessential regions, the coat proteins are able to display foreign peptides without loss of infectivity, and can accumulate in high concentration in the infected bacterial hosts, as summarized by Kehoe and Kay [21]. Most commonly used coat proteins are pIII and pVIII as C-terminus or N-terminus fusions with target peptide library. However, N-terminal fusions to pIII are more tolerant to large insertions.

Random peptide libraries expressed as N-terminal pIII fusions are the most commonly used phage display libraries. The length of peptide can vary from 6 to 43 amino acids when expressed as N-terminal fusion [22]. Structural constraints on the peptides can also be imposed to obtain stringent sequence specificity by flanking the peptide sequence by a pair of cysteine residues [23, 24]. Commercial phage display libraries are available from various vendors including New England Biolabs, Create Biolabs, EMD Millipore, and Lucigen. Custom peptide libraries can also be made and practical tips for their construction have been described [25].

In this chapter, an adaptation of phage display using a commercial M13 library is described. This adaptation has been used to identify a novel protein–protein interaction in the Arabidopsis polyadenylation complex [20] and also a larger library of peptides that interacts with other Arabidopsis polyadenylation factor subunits (B. Addepalli, S. Rao, K. P. Forbes et al., unpublished results).

2   Materials

Use deionized water (a sensitivity of 18 MΩ cm at 25 °C) and analytical grade reagents to prepare all solutions. Autoclave and store all reagents at room temperature (unless indicated otherwise). Conform to all waste disposal and safety regulations while preparing and disposing the materials.

1. *E. coli*—strain Rosetta™ 2(DE3) for expression of recombinant proteins

2. Lysis buffer: 50 mM Tris–HCl (pH 7.5), 150 mM NaCl, 1 mM EDTA, 1 mM phenylmethylsulfonyl fluoride (PMSF).

3. High salt wash buffer: 50 mM Tris–HCl (pH 7.5), 2 M NaCl, 1 mM EDTA.

4. Glutathione elution buffer (GEB): 20 mM reduced glutathione, 50 mM Tris–HCl (pH 8.0)

5. Maltose elution buffer (MEB): 20 mM maltose, 50 mM Tris–HCl (pH 7.5)

6. LB medium: Dissolve 10 g Bacto Tryptone, 5 g of yeast extract, and 10 g of NaCl in water and adjust the volume to 1 L. Autoclave in narrow mouth flask or bottle and store at room temperature.

7. IPTG-Xgal mix: Mix 1.25 g IPTG (isopropyl β-D-thiogalactoside) and Xgal (5-bromo-4-chloro-3-indolyl-β-D-galactoside) in 25 mL dimethyl formamide (DMF). Store the solution in dark at –20 °C.

8. LB-IPTG-Xgal plates: LB medium + 15 g/L agar (either technical or Bacto grade). Autoclave, cool to <70 °C, add 1 mL IPTG/Xgal and pour into petri plates. Store plates at 4 °C in the dark.

9. Agarose top: Dissolve 10 g Bacto Tryptone, 5 g yeast extract, 5 g NaCl, 1 g MgCl$_2$·6H$_2$O per 1 L water and add 7 g agarose. Autoclave, dispense into aliquots of 100 mL. Store as solid at room temperature, melt the medium as needed.

10. Tetracycline solution: Dissolve 20 mg per mL of ethanol. Store in dark at –20 °C. Vortex before use.

11. LB-Tet or appropriate antibiotic Plates: LB medium + 15 g/L agar (either technical or Bacto grade). Autoclave, cool to

<70 °C, add 1 mL Tetracycline solution after vortexing and pour into petri plates. Store plates at 4 °C in the dark.

12. Plate binding buffer: 0.1 M $NaHCO_3$ (pH 8.6).

13. Blocking buffer: 0.1 M $NaHCO_3$ (pH 8.6), 5 mg/mL BSA, 0.02 % NaN_3. Filter sterilize, store at 4 °C.

14. TBS: 50 mM Tris–HCl (pH 7.5), 150 mM NaCl. This can also be prepared as 10× solution. Autoclave and store at room temperature.

15. TBST: TBS + 0.1 % Tween-20 (v/v)

16. PEG/NaCl: 20 % polyethylene glycol-8000, 2.5 M NaCl. Autoclave, store at room temperature.

17. Phage elution solution: 0.2 M Glycine-HCl (pH 2.2), 1 mg/mL bovine serum albumin (BSA)

18. Neutralization solution: 1 M Tris–HCl (pH 9.1)

19. Agarose, molecular biology grade

20. TBE: 45 mM Tris-borate, pH 8.0, 1 mM Na-EDTA.

---

3 Methods

3.1 Preparation of Bait Proteins (See Note 1)

1. Grow 10 mL of an overnight culture of transformed Rosetta cells with the appropriate recombinant plasmid (*see* **Notes 1** and **2**, refer chapter 9 for more details).

2. Use 1–3 mL of the above overnight culture to inoculate 200-mL to 1 L of LB broth media with appropriate antibiotic amounts and grow at 37 °C for 3–5 h. Draw 1 mL cultures to measure the OD at 600 nm between these time periods.

3. Once the OD_{600} reaches 0.7–0.8, induce the fusion protein production by adding 120–200 μL of 1 M IPTG so as to obtain 0.6–1 mM final concentration for 200 mL culture. Optimal induction time and temperatures are determined empirically for each protein. Typically 2–3 h of induction at temperatures between 25 and 37 °C is sufficient.

4. Following induction, harvest cells by centrifugation for 5 min at $6,000 \times g$ and resuspend in 5 mL of lysis buffer. The cells may be frozen at −80 °C for long-term storage at this step.

5. Disrupt cells by sonication (three bursts, 30 s each in a cold room) and remove debris by centrifugation ($12,000 \times g$/15 min). Cells can also be gently lysed with lysozyme (2 mg/mL) at 4 °C before centrifugation.

6. Pass the lysate through 0.42 μm filter to remove particulates (*see* **Note 3**).

7. Pack 1–2 mL bed volumes of the appropriate resin (glutathione-Sepharose for GST, amylose for MBP) in a disposable column

(similar to poly-prep columns #731-1550EDU, Bio-Rad) and equilibrate with 10 volumes of lysis buffer.

8. Pass the filtered extracts through the resin to allow the binding of protein to the column matrix; the flow-through may be passed through column, once again to maximize yield. Subsequently, wash the column with 5 mL lysis initially, followed by 15 mL of wash buffer and 10 mL of lysis buffer (*see* **Note 4**).

9. Elute the fusion proteins with 1–3 mL of appropriate buffer (GEB for GST-fusion, MEB for MBP-fusion proteins). Collect fractions of 250–500 μL and assess by measuring absorbance at 280 nm.

10. Dialyze purified fusion proteins at 4 °C for 12 h with the suitable tag cleavage buffer (*see* **Note 5**), if desired, or plate binding buffer if the protein is going to be used directly for screening phage library. (*See* **Note 6** for further comments regarding protein purity.)

3.2 Phage Titering (See Note 7)

1. A single colony of a suitable *E. coli* host strain (*see* **Note 8**) is used to inoculate 10 mL of LB and incubate at 37 °C with vigorous shaking until mid-log phase ($OD_{600} \sim 0.5$)

2. Melt agarose top in microwave and dispense 3 mL (one per each phage dilution) into sterile culture tubes. The agarose can be kept in melted condition at 45 °C until used.

3. LB/IPTG/Xgal plates used for plating phage (one for each dilution) are pre-warmed at 37 °C to enable fast growth of host bacterium.

4. Usually tenfold serial dilutions of phage are performed in LB. The preferred dilution ranges include 10^8–10^{11} for amplified phage culture supernatants, and 10^1–10^4 for unamplified panning eluate. Always make it a point to use a fresh pipet tip for each dilution, and also use aerosol-resistant pipet tips to avoid cross contamination.

5. Once the host strain, inoculated in LB (**step 1**), reaches mid-log phase, dispense 200 μL culture in a microfuge tube for each phage dilution.

6. To each dispensed culture, add 10 μL of diluted phage, vortex quickly, and incubate at room temperature for 5 min.

7. Transfer the infected cells to the top agarose in a culture tube kept at 45 °C, quickly vortex the contents and pour onto the pre-warmed LB/IPTG/Xgal plate, and allow it to spread evenly by tilting the plate.

8. Cool the contents of plate for 10 min to solidify the agarose and then keep the plates inverted overnight at 37 °C.

9. Count the blue plaques on each plate especially those having about $\sim 10^2$ plaques. Phage titer is obtained by multiplying each number with dilution factor for the given plate and is referred to as plaque forming units (pfu) per 10 μL.

3.3 Phage Panning

1. Bring the purified bait protein (Subheading 3.1) to a concentration of 0.1 µg/µL in plate binding buffer. Use this to coat individual wells of a sterile polystyrene 12-well plates; for this, add about 700 µL of protein solution to each well and swirl repeatedly to wet the surface completely (*see* **Note 9**).

2. Incubate the plate with protein solution overnight at 4 °C with gentle agitation in a humidified container (wrapped with wet paper towels in a zipped polybag). The plates can be stored at 4 °C in this container until ready for panning.

3. Carefully and quickly pour off the solution and invert the plate onto a clean paper towel to remove any residual droplets (*see* **Note 10**).

4. Turn the plates "right-side up" and fill each well with blocking buffer. Incubate at 4 °C for 1 h.

5. Carefully and quickly discard the blocking buffer and invert the plate onto a clean paper towel to remove any residual droplets (*see* **Note 10**).

6. Re-invert the plate and wash each well by coating the bottom and sides of well with TBST and swirling gently. Discard the wash solution and invert the plate onto a clean paper towel to remove any residual droplets (*see* **Note 10**). Quickly proceed to the next step to avoid drying of well interior.

7. For each well that has been coated with bait protein, dilute approximately 2×10^{11} phage (10 µL of original PhD kit library if it is from New England Biolabs) into 700 µL TBST (*see* **Note 11**). Add this to the coated well and rock gently at room temperature for 1 h.

8. After 1 h of binding, carefully and quickly pour off the solution and invert the plate onto a clean paper towel to remove any residual droplets (*see* **Note 10**). Re-invert the plate and wash each well by coating the bottom and sides of well with TBST and swirling. Discard the wash solution and invert the plate onto a clean paper towel to remove any residual droplets (*see* **Note 10**).

9. Elute the bound phage by adding 700 µL of phage elution solution and rocking gently for <10 min. Transfer the eluted solution to microfuge tube and neutralize with 105 µL of neutralization solution.

10. Amplify the eluted phage by adding the eluate to 20 mL of a log-phase culture of the selected phage host strain (*see* **Note 8**). Incubate at 37 °C for 4.5 h with vigorous shaking.

11. Transfer the culture at the end of incubation to a centrifuge tube and centrifuge for 10 min at $10,000 \times g$. Recover the supernatant to a new centrifuge tube and repeat the centrifugation for 10 min at $10,000 \times g$.

12. Transfer the top 80 % of the supernatant to a new tube and add 1/6 volume of PEG/NaCl. Incubate at 4 °C for 1–18 h.

13. Centrifuge the precipitated phage at 4 °C for 15 min at 10,000 × g. Decant the supernatant and briefly centrifuge the tube. Remove residual supernatant with a small pipette.

14. Resuspend the pellet in 1 mL of TBS and centrifuge the suspension at 6,000 × g for 5 min at 4 °C to remove the residual cells.

15. Transfer the supernatant to a microfuge tube and re-precipitate the phage with 1/6 volume of PEG/NaCl. After incubation on ice for 45–60 min, centrifuge for 10 min at 10,000 × g. Decant the supernatant and remove the last traces of residual supernatant with a micropipette.

16. Resuspend the pellet in 200 μL TBS and add NaN_3 to a final concentration of 0.02 %. Remove insoluble material with a brief spin at top speed in a microcentrifuge.

17. Estimate the phage titer as described in Subheading 3.2.

18. Repeat **steps 1–17** of this section until the desired extent of enrichment is achieved (*see* **Note 12**).

19. Determine the titer of final eluate; these plaques may directly be used for sequencing (*see* **Note 13**). Store the eluate at 4 °C.

3.4 Characterization of Affinity-Purified Phage

1. If it has not been done, amplify the purified phage as described in Subheading 3.3, **step 10**. For this, dilute an overnight culture of the *E. coli* host (*see* **Note 8**) by 100-fold in LB, and add 10 μL of unamplified eluate phage to 1 mL of the diluted culture. Incubate at 37 °C with shaking for 4.5–5 h.

2. Transfer cultures to a microcentrifuge tube and pellet the cells for 30 s at top speed in a microcentrifuge. Transfer 80 % of the supernatant (avoiding the cell pellet) to a fresh microcentrifuge tube. This is the amplified stock (*see* **Note 14**).

3. The resulting library present in this amplified stock may be characterized by high-throughput DNA sequencing (*see* **Note 15**). For this, amplify 1 μL of the stock in a 25 μL PCR reaction using primers that are designed to amplify that portion of the phage genome that includes the inserted foreign DNA sequences (Fig. 1; *see* **Note 16**).

4. Separate the PCR products on an agarose gel cast and run in TBE (*see* **Note 17**). Gel-purify the amplification products using an appropriate kit (*see* **Note 18**). For example, follow the instructions (exactly) for the Qiagen Gel Purification System, and elute the purified product into 50 μL of the supplier's Elution Buffer.

5. Add 1 μL of the purified PCR product into a 25 μL PCR reaction containing primers designed to append high throughput

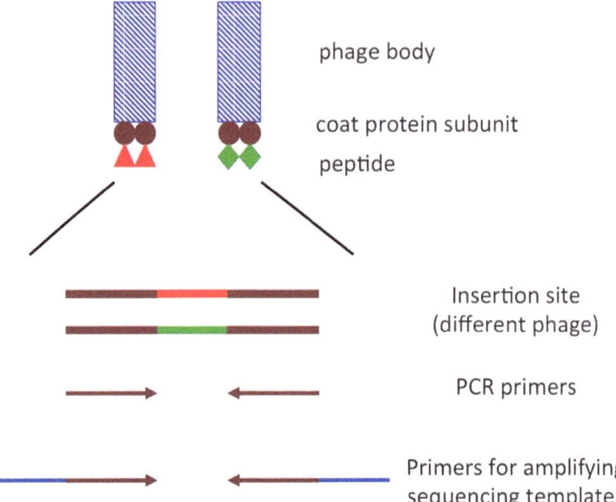

Fig. 1 Illustration of the generic primer design for characterization of affinity-purified phage. At the *top* is a cartoon of the structure of a filamentous phage (such as M13). The coat protein subunit around which a combinatorial library is built (such as the pIII subunit of M13) is shown in *brown*. Individual peptides attached to the coat protein subunit are represented as *red* or *green symbols*. Beneath the illustration of the phage is a depiction of the DNA sequences surrounding the location of those encoding the different peptides; color coding corresponds to that for the proteins shown above. PCR primers derived from the coat protein sequences are represented as *arrows* (*see* **Note 16**), and primers used to adapt PCR products for high throughput sequencing as *hybrid arrows*, with the *blue* segments representing the sequencing platform-specific adapter sequences (color figure online)

sequencing adapters onto the PCR products (*see* **Note 16**). Amplify these products for a minimal number of cycles (5–15, depending on the concentration of PCR product obtained in **step 4**).

6. Separate the PCR products on an agarose gel cast and run in TBE (*see* **Note 17**). Gel-purify the amplification products using an appropriate kit (*see* **Note 18**). For example, follow the instructions (exactly) for the Qiagen Gel Purification System, and elute the purified product into 50 μL of the supplier's Elution Buffer.

7. Submit the sequencing samples to the facility of one's choice (*see* **Notes 15** and **19**).

8. Assess the returned peptide sequences using BLAST (locally or at NCBI). For this, parameters must be set so as to permit analysis of short sequences; typically, this requires that the Expect threshold be set for a maximum value (1,000), and word size adjusted to recover the largest numbers of hits.

4   Notes

1. The availability of highly pure bait proteins is critical to the successful identification of affinity peptides that might resemble the contact points in the interacting partners. We express the target proteins in bacterial host as fusion proteins with affinity tags. Ideally, the tag needs to be cleaved to remove its interference in the screening process, followed by purification of bait protein by chromatography or other biochemical methods. In case the tag cannot be specifically digested, alternative procedure involving prior incubation of phage library with the affinity tag may be employed to remove the tag specific peptides. A generic protocol is described in chapter 9 for purification of bait proteins. To produce the required fusion bait proteins, clone the appropriate gene sequence in the desired expression vector where the expression is driven by the inducible promoter. The most common tags used in the fusion protein production are glutathione-S-transferase (pGEX-2T, Pharmacia), maltose binding protein (MBP, pMAL), his-tag (Novagen), calmodulin-binding protein (Stratagene). Tags like MBP have chaperone-like activity that helps in folding newly synthesized foreign proteins while smaller tags do not provide the same kind of advantage.

2. The recombinant plasmids that encode tagged fusion proteins are transformed into Rosetta™ 2(DE3) cells for protein production. These cells supply tRNAs for translating rare codons to enable improved expression of eukaryotic proteins.

3. Filtration helps in maintaining uniform flow in subsequent chromatographic steps without clogging of the columns.

4. The wash solutions may be screened for the presence of target bait protein by SDS-PAGE, if needed. This helps to determine the optimal procedure for different proteins.

5. If desired, the affinity tag (GST or MBP) may be cleaved using a suitable enzyme (thrombin/thrombin agarose in case of GST-fusion, factor Xa for MBP as per the manufacturer's recommendations) and subjected to second round of affinity chromatography to remove the affinity tag.

6. The purity of bait protein, especially those from which the tag has been cleaved, may further be enhanced by ion-exchange chromatography. While the details will vary depending on the nature of the bait (and are thus not described here), the general approach will entail selection of a suitable column (anion or cation exchange), and purification of the bait with a linear gradient (0.025–2 M) of NaCl in lysis buffer. Protein-containing fractions are typically identified by monitoring absorbance at 280 nm, and protein purity is confirmed by

SDS-PAGE. Fractions with pure bait protein are pooled and dialyzed against plate binding buffer.

7. It is necessary to estimate the phage titer for the purchased or prepared display library, and to proceed with the rounds of enrichment for interacting phage. This keeps the multiplicity of infection (MOI) less than 1. A low MOI is needed to assure that each plaque contains only one DNA sequence.

8. A robust F$^+$ strain (commercial strains like ER2738 from New England Biolabs https://www.neb.com/products/e8100-phd-7-phage-display-peptide-library-kit) with a rapid growth rate is ideally suited for M13 propagation. The host bacterium needs to be plated on media so that it is selected for the F factor. Additionally, single colonies, rather than a glycerol stock, are to be used for inoculation of medium for phage amplification and maintenance.

9. Sufficient wetting of the plate is necessary to allow optimal protein binding to polystyrene plate.

10. Speed at this step is important. Care must be taken to avoid repeated inversion and other handling of the plate beyond the initial inversion and setting on the paper towel.

11. If fusion protein is going to be directly used for panning, then the phage is initially incubated with tag protein (GST, MBP or other appropriate protein) for 15 min before adding to the protein coated polystyrene plates.

12. Successive rounds of panning can be carried out with 10^9 pfu of input phage at **step** 7 if the titer of the current eluate phage is too low. Typically, three rounds of panning suffice to permit identification of interacting peptides.

13. Plates with plaques to be used for sequencing should not be incubated at 37 °C for greater than 18 h; longer period lead to deletions and other loss of phage.

14. The short-term storage for this is 4–5 weeks at 4 °C. The stock should be diluted with sterile glycerol 1:1 for long-term storage at –20 °C.

15. By adapting the amplified library for high-throughput DNA sequencing, the time-consuming and costly process of characterizing individual phage clones is eliminated. The population of inserts in the purified phage may be adapted for sequencing using any of a number of current platforms, by re-amplifying the PCR products for 10–15 cycles using hybrid primers that consist of platform-specific adapters flanking the phage library-specific sequences. This strategy is illustrated in Fig. 1. This approach is readily adapted for sequencing on the 454, Ion Torrent, and MiSeq instruments. Importantly, only a few thousand single-end sequence reads, of 100 nts (or fewer)

each, are all that is needed to fully characterize the enriched phage library.

16. PCR reactions utilize standard protocols and enzymes from many sources. "Standard" as well as hot-start enzymes may be used. Primers are designed to amplify the insert. For M13 pIII libraries (as described here), the primers are 5′-AAAGCCTCTGTAGCCGTTGC-3′ (forward primer) and 5′-AGCCACCACCCTCATTTTCA-3′ (reverse primer). To adapt PCR products for high throughput sequencing, the appropriate sequencing adapters are added to the 5′-ends of these forward and reverse primers, as illustrated in Fig. 1. Specific sequences are not provided here, but may be obtained from the sequencing provider or core facility. (Different sequences will be used for the most common platforms— Miseq, Ion Torrent, and 454.)

17. The agarose concentration for these gels will be dictated by the size of the PCR products of interest. Typically, 1 % (w/v)–2 % (w/v) gels are used.

18. Gel-purification of PCR products may be performed using the user's choice of kit and methods. Each approach will entail kit-specific steps and components, and are thus not described here. The end goal for this is the purification of sufficient PCR product to permit sequencing (*see* **Notes 15** and **18**).

19. The most cost-effective approach for sequencing enriched libraries will entail the return of 10^5–10^6 individual sequences, depending on the sequencing technology. Multiple experiments can easily be combined into a single sequencing run by incorporating experiment-specific bar codes into the primers used in **step 5** of Subheading 3.4; these bar codes are placed immediately upstream of the phage-specific sequences of the primers used to amplify sequencing templates (Fig. 1), such that they are the first bases read in the output. In all cases, sequences may be analyzed and resolved into contigs that correspond to individual high-affinity interacting peptides using the user's choice of standard read-mapping and assembly software packages. Importantly, the numbers of sequences and sizes of data files make projects such as this amenable to relatively inexpensive commercially available software packages such as Geneious (Biomatters Ltd.).

Acknowledgements

This work was supported by the US National Science Foundation (MCB-0313472 and IOS-0817818).

References

1. Zhao J, Hyman L, Moore C (1999) Formation of mRNA 3′ ends in eukaryotes: mechanism, regulation, and interrelationships with other steps in mRNA synthesis. Microbiol Mol Biol Rev 63(2):405–445

2. Gilmartin GM (2005) Eukaryotic mRNA 3′ processing: a common means to different ends. Genes Dev 19(21):2517–2521. doi:10.1101/gad.1378105

3. Hunt AG, Xu R, Addepalli B, Rao S, Forbes KP, Meeks LR, Xing D, Mo M, Zhao H, Bandyopadhyay A, Dampanaboina L, Marion A, Von Lanken C, Li QQ (2008) Arabidopsis mRNA polyadenylation machinery: comprehensive analysis of protein-protein interactions and gene expression profiling. BMC Genomics 9:220. doi:10.1186/1471-2164-9-220

4. Hunt AG, Xing D, Li QQ (2012) Plant polyadenylation factors: conservation and variety in the polyadenylation complex in plants. BMC Genomics 13:641. doi:10.1186/1471-2164-13-641

5. Minvielle-Sebastia L, Keller W (1999) mRNA polyadenylation and its coupling to other RNA processing reactions and to transcription. Curr Opin Cell Biol 11(3):352–357. doi:10.1016/S0955-0674(99)80049-0

6. Moore MJ, Proudfoot NJ (2009) Pre-mRNA processing reaches back to transcription and ahead to translation. Cell 136(4):688–700. doi:10.1016/j.cell.2009.02.001

7. Smith GP (1985) Filamentous fusion phage: novel expression vectors that display cloned antigens on the virion surface. Science 228(4705):1315–1317

8. Smith WC, Hargrave PA (2000) Mapping interaction sites between rhodopsin and arrestin by phage display and synthetic peptides. Methods Enzymol 315:437–455

9. Sidhu SS, Fairbrother WJ, Deshayes K (2003) Exploring protein-protein interactions with phage display. Chembiochem 4(1):14–25. doi:10.1002/cbic.200390008

10. Jestin JL (2008) Functional cloning by phage display. Biochimie 90(9):1273–1278. doi:10.1016/j.biochi.2008.04.003

11. Mersich C, Jungbauer A (2008) Generation of bioactive peptides by biological libraries. J Chromatogr B Analyt Technol Biomed Life Sci 861(2):160–170. doi:10.1016/j.jchromb.2007.06.031

12. Chen B, Cao S, Zhang Y, Wang X, Liu J, Hui X, Wan Y, Du W, Wang L, Wu K, Fan D (2009) A novel peptide (GX1) homing to gastric cancer vasculature inhibits angiogenesis and cooperates with TNF alpha in anti-tumor therapy. BMC Cell Biol 10:63. doi:10.1186/1471-2121-10-63

13. Pande J, Szewczyk MM, Grover AK (2010) Phage display: concept, innovations, applications and future. Biotechnol Adv 28(6):849–858. doi:10.1016/j.biotechadv.2010.07.004

14. Georgieva Y, Konthur Z (2011) Design and screening of M13 phage display cDNA libraries. Molecules 16(2):1667–1681. doi:10.3390/molecules16021667

15. Li W (2012) ORF phage display to identify cellular proteins with different functions. Methods 58(1):2–9. doi:10.1016/j.ymeth.2012.07.013

16. Liu BA, Engelmann BW, Nash PD (2012) High-throughput analysis of peptide-binding modules. Proteomics 12(10):1527–1546. doi:10.1002/pmic.201100599

17. Van Dorst B, Mehta J, Rouah-Martin E, Blust R, Robbens J (2012) Phage display as a method for discovering cellular targets of small molecules. Methods 58(1):56–61. doi:10.1016/j.ymeth.2012.07.011

18. Ducki S, Bennett E (2009) Protein-protein interactions: recent progress in the development of selective PDZ inhibitors. Curr Chem Biol 3(2):146–158

19. Petschnigg J, Snider J, Stagljar I (2011) Interactive proteomics research technologies: recent applications and advances. Curr Opin Biotechnol 22(1):50–58. doi:10.1016/j.copbio.2010.09.001

20. Addepalli B, Hunt AG (2008) The interaction between two Arabidopsis polyadenylation factor subunits involves an evolutionarily-conserved motif and has implications for the assembly and function of the polyadenylation complex. Protein Pept Lett 15(1):76–88

21. Kehoe JW, Kay BK (2005) Filamentous phage display in the new millennium. Chem Rev 105(11):4056–4072. doi:10.1021/cr000261r

22. McConnell SJ, Uveges AJ, Fowlkes DM, Spinella DG (1996) Construction and screening of M13 phage libraries displaying long random peptides. Mol Divers 1(3):165–176

23. McLafferty MA, Kent RB, Ladner RC, Markland W (1993) M13 bacteriophage displaying disulfide-constrained microproteins. Gene 128(1):29–36

24. Clackson T, Wells JA (1994) In vitro selection from protein and peptide libraries. Trends Biotechnol 12(5):173–184. doi:10.1016/0167-7799(94)90079-5

25. Fukunaga K, Taki M (2012) Practical tips for construction of custom Peptide libraries and affinity selection by using commercially available phage display cloning systems. J Nucleic Acids 2012:295719. doi:10.1155/2012/295719

Chapter 14

Genome-Wide Determination of Poly(A) Site Choice in Plants

Pratap Kumar Pati, Liuyin Ma, and Arthur G. Hunt

Abstract

Second generation DNA sequencing technologies have been a great boon for the study of mRNA polyadenylation. The experimental determination of large numbers of polyadenylation sites using high-throughput sequencing strategies has provided the necessary platform for deeper understanding the regulation of gene expression in eukaryotes. For generating large sets of data to map poly-A sites, specialized sample preparations that target the junction of 3′-UTR and the poly(A) tail are usually employed. Here, we describe three different protocols that are effectively used for global determinations of poly(A) site choice in plants.

Key words Alternative polyadenylation, High-throughput sequencing, Gene expression, PAT-seq

1 Introduction

Polyadenylation of mRNA is a necessary step in the expression of eukaryotic genes and for mRNA metabolism. Additional knowledge in this area clearly indicates the important role of alternative polyadenylation (APA) in numerous processes including oncogene regulation [1, 2], development [3–6], and cellular differentiation [7, 8]. In plants, the role of APA in the regulation of flowering time [9–13], oxidative stress responses [14, 15], and the expression of genes involved in RNA processing [16] has been reported. With the refinement of experimental tools and availability of important biological resources, additional biological processes linked to APA will be unraveled in future.

The advent of the second generation sequencing technology has greatly facilitated the study of gene expression. In particular, RNA-seq has been very efficiently used to characterize transcriptomes. In a regular RNA-seq protocol, only a small proportion of poly(A) sites are queried due to the randomness of RNA fragmentation. This makes difficult a comprehensive study of poly(A) sites on a genome-wide basis. For high-throughput determinations of

Arthur G. Hunt and Qingshun Quinn Li (eds.), *Polyadenylation in Plants: Methods and Protocols*, Methods in Molecular Biology, vol. 1255, DOI 10.1007/978-1-4939-2175-1_14, © Springer Science+Business Media New York 2015

poly(A) site, a number of *poly(A) tag sequencing* (PAT-seq) protocols have been developed that target only the junctions of the 3′-UTR and poly(A) tails [17–20]. Here, we describe three variations of PAT-seq protocols which are effectively utilized for gene expression analysis as well as for global determinations of poly(A) site choice in plants. In Method 1 (Fig. 1), restriction enzymes are used to digest the cDNA before ligating it to adapter and followed by sequencing whereas, in Methods 2 and 3, RNA fragmentation is introduced before reverse transcription. Steps in Method 2 and Method 3 diverge after the preparation of cDNA (Fig. 1). These protocols are versatile and reproducible [21] and may be used to study genome-wide poly(A) site choice (e.g., [14, 20]) as well as overall gene expression levels [21].

2   Materials

Prepare all solutions using nuclease-free ultrapure water (18 MΩ cm at 25 °C). All reagents used are analytical grade. Prepare and store all reagents at room temperature (unless indicated otherwise). All waste disposal regulations are strictly followed when disposing waste materials.

2.1   RNA Isolation

1. Trizol reagent.

2. Liquid nitrogen.

3. Chloroform.

4. Isopropyl alcohol.

5. Pine Tree extraction buffer (*see* **Note 1**): 2 % cetyltrimethyl ammonium bromide (CTAB), 2 % polyvinylpyrrolidone (K30), 100 mM Tris–HCl (pH 8.0), 25 mM EDTA, 2.0 M NaCl, 0.5 g/L spermidine, 2 % beta-mercaptoethanol. (Heat the extraction buffer to dissolve all components except the beta-mercaptoethanol and mix well. Aliquot 5 mL into the 15 mL tubes and add the BME just prior to use.)

6. Chloroform/isoamyl alcohol: mix 24 volumes of chloroform and 1 volume of isoamyl alcohol.

7. SSTE: 1.0 M NaCl, 0.5 % SDS, 10 mM Tris–HCl (pH 8.0), 1 mM EDTA (pH 8.0).

8. Ethanol, 100 %.

9. High speed centrifuge.

10. Ethanol, 70 % (v/v).

11. RNAse-free DNase I and 10× DNAse I reaction buffer.

12. RNeasy Plant Mini Kit (Qiagen)—this kit contains columns and the RLT and RPE solutions used in purifying RNA.

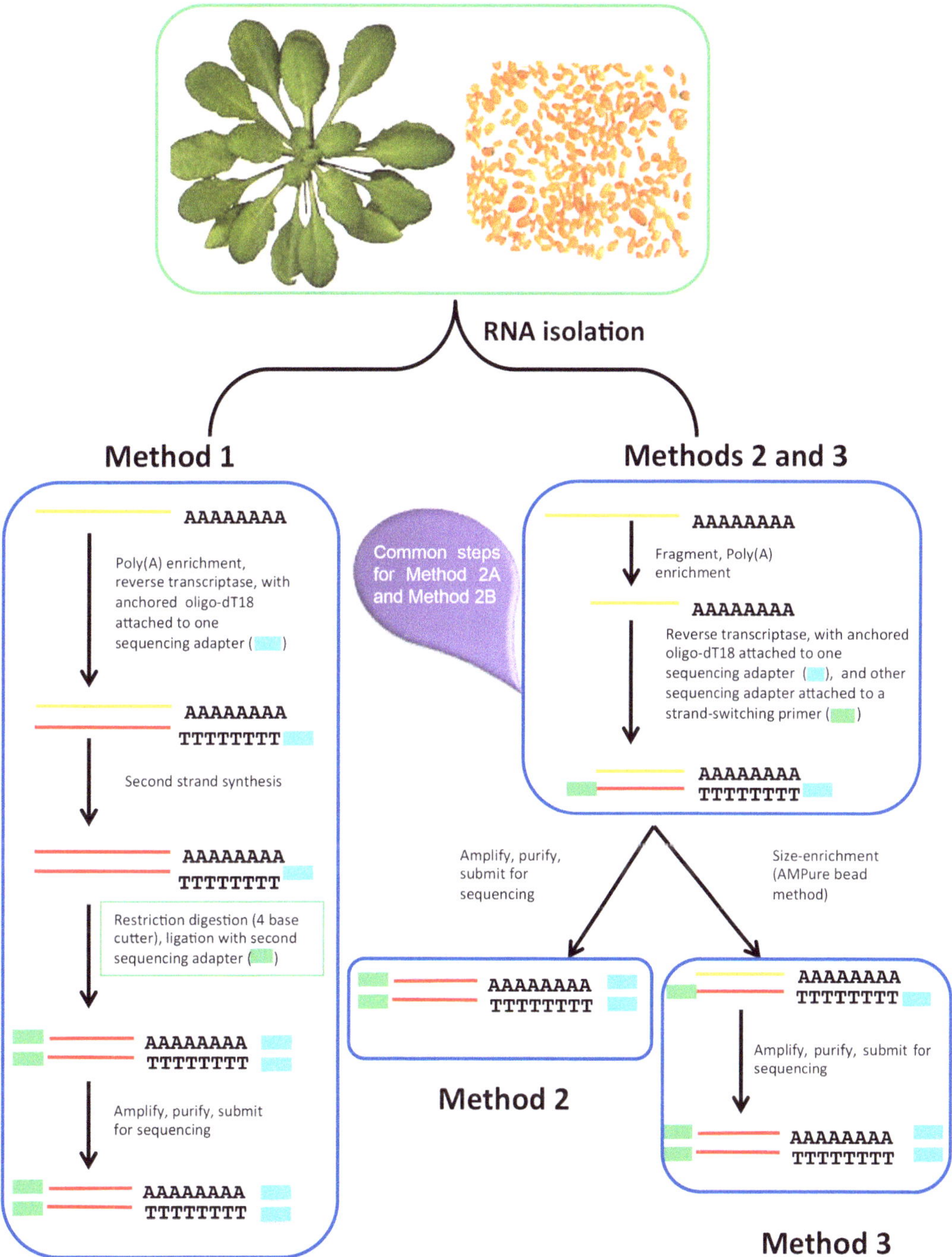

Fig. 1 Illustration of the three methods of PAT-Seq that are described in this chapter

13. NanoDrop Instrument.

14. Microcentrifuge.

15. Bioanalyzer (Agilent Technologies; Model 2100).

2.2 Preparation of Poly(A) Tags (PATs)

1. Oligo-dT magnetic beads.

2. Magnetic separation rack.

3. Binding buffer for poly(A) enrichment: 20 mM Tris–HCl pH 7.5, 1 M LiCl, 2 mM Na-EDTA.

4. Wash buffer for poly(A) enrichment: 10 mM Tris–HCl pH 7.5, 0.15 M LiCl, 1 mM Na-EDTA.

5. Poly(A) elution buffer for poly(A) enrichment: 10 mM Tris–HCl pH 7.5.

6. DNA primers and adapters (Table 1).

7. QIAquick PCR Purification Kit (includes spin columns and buffers for column purification, washing, and elution of PCR products; Qiagen).

8. Restriction endonucleases *Nla* III and *Tai* I.

9. RNA fragmentation buffer: 100 mM $ZnCl_2$ in 100 mM Tris–HCl, pH 7.0.

10. RNA fragmentation stop solution: 0.1 M Na-EDTA, pH 8.0.

11. SMARTScribe™ 5× first strand buffer and SMARTScribe™ Reverse Transcriptase (Clontech).

12. 10× deoxynucleotide triphosphate stock solution (dNTPs) for reverse transcription: 10 mM each of dATP, dTTP, dCTP, and dGTP.

13. 0.1 M dithiothreitol.

14. RNase inhibitor, human placental (1 U/μL).

15. RNase H, 5 U/μL.

16. RNase A/T1 Mix—2 mg/mL RNAse A, 5,000 U/mL RNAse T1 in 50 mM Tris–HCl, pH 7.4 + 50 % (v/v) glycerol.

17. Phire Hot Start II DNA Polymerase and 5× buffer (Thermo Scientific).

18. 10× PCR dNTP mix: 2.5 mM each of dATP, dTTP, dCTP, and dGTP.

19. Programmable thermocycler with heated lid.

20. 10 mM Tris–HCl, pH 7.5 + 0.1 mM Na-EDTA.

21. T4 DNA ligase and 10× ligation buffer.

22. 10 mM Tris–HCl, pH 7.5.

23. Agencourt® Ampure®XP beads (Beckman Coulter, Inc.).

24. S1 Nuclease and 5× S1 nuclease reaction buffer.

25. Agarose, molecular biology grade.

26. TBE: 45 mM Tris-borate, pH 8.0, 1 mM Na-EDTA.

27. Apparatus for agarose gel electrophoresis (gel box and casting tray, power supply).

Table 1
Primers and sample keys (bar codes are underlined)

Designation	Sequence (5′ → 3′)	Use
RT-PE1 series	5′-biotin-TCCTGCTGAACCGCTCTTCCGATCT NNN<u>XX</u>TTTTTTTTTTTTTTTTTTTTVN	Reverse transcription primer (Method 1)
RT-PE2 series	TCCTGCTGAACCGCTCTTCCGATCTNNN<u>XX</u> TTTTTTTTTTTTTTTTTVN	Reverse transcription primer (Method 1)
PE-RTbio	5′-biotin–TCCTGCTGAACCGCTCTTCCGATCT	Second strand synthesis (Method 1)
SWITCH1.1	AAGC<u>ACGT</u>CATGAACGCAGAGTGGCCAGGG	Strand switching (Method 1)
SWITCH1.2	AAGC<u>ACGT</u>CATGAACGCAGAGTGGC	Second strand synthesis (Method 1)
PE-ADN1A	ACACTCTTTCCCTACACGACGCTCTTCCGAT CTCATG	Ligation of 5′ Illumina adapter (Method 1)
PE-ADT1A	ACACTCTTTCCCTACACGACGCTCTTCCGAT CT<u>ACGT</u>	Ligation of 5′ Illumina adapter (Method 1)
PE-AD2A	AGATCGGAAGAGCGTCGTGTAGTCTGT CGAGAC	Ligation of 5′ Illumina adapter (Method 1)
PE-ADN1B	ACACTCTTTCCCTACACGACGCTCTTCCGAT CT<u>GA</u>CATG	Ligation of 5′ Illumina adapter (Method 1)
PE-ADT1B	ACACTCTTTCCCTACACGACGCTCTTCCGAT CT<u>GAACGT</u>	Ligation of 5′ Illumina adapter (Method 1)
PE-AD2B	TCAGATCGGAAGAGCGTCGTGTAGTCTGT CGAGAC	Ligation of 5′ Illumina adapter (Method 1)
PE-ADN1C	ACACTCTTTCCCTACACGACGCTCTTCCG ATCT<u>CT</u>CATG	Ligation of 5′ Illumina adapter (Method 1)
PE-ADT1C	ACACTCTTTCCCTACACGACGCTCTTCCGA TCT<u>CTACGT</u>	Ligation of 5′ Illumina adapter (Method 1)
PE-AD2C	AGAGATCGGAAGAGCGTCGTGTAGTC TGTCGAGAC	Ligation of 5′ Illumina adapter (Method 1)
PE-ADN1D	ACACTCTTTCCCTACACGACGCTCTTCC GATCT<u>CC</u>CATG	Ligation of 5′ Illumina adapter (Method 1)
PE-ADT1D	ACACTCTTTCCCTACACGACGCTCTTCC GATCT<u>CCACGT</u>	Ligation of 5′ Illumina adapter (Method 1)
PE-AD2D	GGAGATCGGAAGAGCGTCGTGTAGTC TGTCGAGAC	Ligation of 5′ Illumina adapter (Method 1)
RT-PE3 series	ACACTCTTTCCCTACACGACGCTCTTCCG ATCTNN<u>XXX</u>TTTTTTTTTTTTTTTTTTTTVN	Reverse transcription primer (Methods 2 and 3)
SWITCH2	CGGTCTCGGCATTCCTGCTGAACCGCTCT TCCGATCTGGG	Strand switching (Methods 2 and 3)
PE-PCR1	AATGATACGGCGACCACCGAGATCTACACT CTTTCCCTACACGACGCTCTTCCGATCT	Tag amplification (all Methods)
PE-PCR2	CAAGCAGAAGACGGCATACGAGATCGGTCTC GGCATTCCTGCTGAACCGCTCTTCCGATCT	Tag amplification (all Methods)

Nla III sites in SWITCH primers are boxed while *Tai I* sites are double underlined. "N"—random base; "V"—A, C, or G, randomized in the primer population; "<u>XX</u>", "<u>XXX</u>", and other singly underlined bases—positions of bar codes used to distinguish samples for multiplex sequencing

28. Sterile scalpel for dissection of agarose gels.

29. Qiagen Gel Extraction Kit for DNA purification (includes spin columns and buffers for solubilization of agarose gel fragments, column purification, washing, and elution of PCR products; Qiagen).

3 Methods

The series of steps for preparing short cDNA tags for high throughput sequencing is illustrated in Fig. 1 and may be summarized as: isolate RNA (Subheading 3.1), prepare RNA for tag production (Subheading 3.2), and prepare poly(A) tags using one of three methods (Subheading 3.3).

3.1 Isolation of RNA

3.1.1 RNA Isolation from Leaves and Roots

1. Freeze tissue in liquid nitrogen in a mortar, grind to a powder with a pestle. Cool the mortar and pestle prior to use with liquid nitrogen, and work quickly so as to avoid thawing of the frozen tissue.

2. Add 1 mL of Trizol reagent for every 50–100 mg of tissue, homogenize completely using the pestle. Incubate for 5 min at room temperature.

3. Add 0.2 mL chloroform per mL of Trizol. Shake vigorously for 15 s and incubate at room temperature for 3 min. Centrifuge at $12,000 \times g$ for 15 min.

4. Transfer the aqueous to a fresh tube. Add 0.5 mL isopropyl alcohol per mL of Trizol used in **step 1**. Incubate sample at room temperature for at least 10 min (*see* **Note 2**). Centrifuge at $12,000 \times g$ for 10 min.

5. Remove and discard the supernatant, wash the nucleic acid pellet with 1 mL of 75 % EtOH per mL of Trizol in used in **step 1**. Vortex, centrifuge at $7,500 \times g$ for 5 min.

6. Discard supernatant. Air-dry pellet for 5–10 min. Dissolve in RNase-free water.

3.1.2 RNA Isolation from Seeds

1. Cool down mortars with liquid N_2.

2. Grind seed (100–300 mg) in liquid N_2.

3. Transfer powder to a 15 mL centrifuge tube containing 5 mL preheated Pine Tree extraction buffer (an appropriate aliquot of the buffer is preheated to 65 °C prior to use).

4. Vortex to mix well, immediately add 5 mL chloroform/isoamyl alcohol, vortex well (*see* **Note 3**).

5. Centrifuge all samples for 5 min at $3,000 \times g$.

6. Transfer supernatant to fresh sterile tube, add an equal volume of chloroform/isoamyl alcohol, vortex well.

7. Centrifuge for 4 min at $3,000 \times g$, transfer supernatant to fresh sterile tube.

8. Add 0.25 volume 10 M LiCl and store overnight at 4 °C.

9. Centrifuge for 20 min at $12,000 \times g$ at 4 °C and discard supernatant.

10. Dissolve pellet in 500 μL SSTE. Once dissolved, transfer into a 1.5 mL microcentrifuge tube.

11. Add an equal volume of chloroform/isoamyl alcohol, vortex well, spin 2 min at top speed in a microcentrifuge, move supernatant to a fresh microcentrifuge tube.

12. Add 2 volumes 100 % ethanol and incubate for 30 min at −80 °C.

13. Spin 20 min at maximum speed (12,000 Å~ g) in a microcentrifuge at 4 °C.

14. Discard supernatant, wash pellet with 70 % ethanol, spin for 2 min at top speed in a microcentrifuge in the cold room, and discard supernatant.

15. Air-dry pellet, dissolve in 50 μL RNase free H_2O.

16. Determine concentration of RNA on the NanoDrop Instrument.

3.2 Preparation of RNA for Poly(A) Tag Production

1. Bring total RNA to a concentration of 0.1–5 mg/mL with RNAse-free water.

2. Add 10 μL of 10× DNAse I buffer and 2 U of RNAse-free DNAse I to 90 μL of the total RNA. Incubate at 37 °C for 60 min.

3. Add 700 μL of RLT to the reaction, mix, and then add 500 μL of 100 % ethanol (*see* **Note 4.**) Mix and apply to a Qiagen RNeasy column (or an equivalent column from other providers). Centrifuge the column in a microcentrifuge at top speed for 30–90 s, discard the flow-through.

4. Apply 500 μL RPE to the column and the sample centrifuged as above. Discard the flow-through and spin the (empty) column at top speed for 2 min.

5. Move the column to a fresh 1.5 mL tube, add 50 μL RNase-free water, and spin at top speed for 30–60 s.

6. Recover the flow-through, analyze using the NanoDrop instrument to assess the yield of recovery.

3.3 Preparation of Poly(A) Tags (PATs)

Three protocols are described for the production of short cDNA tags that query the mRNA-poly(A) junction. These are subsequently referred to as Method 1, Method 2, and Method 3 (Fig. 1). Methods 2 and Method 3 have common steps up to a point and after which they diverge.

3.3.1 Method 1

In this variation, the ends of cDNA tags that correspond to the 5′ ends of mRNA fragments are generated using restriction enzymes (*see* **Note 5**.)

1. Aliquot 15 μL of oligo-dT magnetic beads into a 200 μL microcentrifuge tube. Wash twice with binding buffer (each wash consists of suspending the beads in 200 μL of buffer, followed by separation of the beads using a magnetic stand). Suspend the beads in 50 μL of binding buffer.

2. Bring 2–20 μg of total RNA to a total volume of 50 μL with RNAse-free water and add to the washed beads from **step 1**. Incubate at room temperature for 5–15 min, then separate the beads using a magnetic stand. Discard the supernatant (*see* **Note 6**).

3. Wash beads twice with 200 μL wash buffer.

4. Suspend the beads in the poly(A) elution buffer and incubate at 80 °C for 3 min. Separate the beads using a magnetic stand and recover the supernatant containing polyadenylated RNA.

5. Mix 14 μL of poly(A)-enriched RNA from **step 4** with 1 μL (containing 100 pmol) of the appropriate RT primer (*see* Table 1) and the SWITCH1.1 primer (*see* Table 1 and **Note 7**). Heat the mixture to 65 °C for 5 min, chill on ice, and repeat the heating and chilling.

6. Add the following (while cold): 5 μL SMARTScribe 5× first strand buffer, 2.5 μL 10× dNTPs for reverse transcription, 1 μL 100 mM DTT, and 0.5 μL RNase Inhibitor. Bring the mixture to 42 °C in a thermocycler that has a heated lid. After 1–2 min (enough time to equilibrate at the selected temperature), add 1 μL (100 U) of SMARTScribe reverse transcriptase. Incubate for 2 h.

7. Heat the reaction to 70 °C for 5 min and then bring to 37 °C. Add 1 μL RNase H + 1 μL RNaseA/T1 (*see* **Note 8**) and incubate 37 °C for 30 min.

8. Purify the cDNA using the QIAquick PCR Purification Kit, following the manufacturer's protocol to the letter. Elute the cDNA in 50 μL of the included elution buffer (EB).

9. Add 2 μL of the purified cDNA to 10 μL of 5× Phire reaction buffer (included with the enzyme), 5 μL of 10× PCR dNTP mix, 20 pmol of the PE-RTbio and SWITCH1.2 primers (*see* Table 1), 0.5 μL of Phire II Hot Start DNA Polymerase, and water to bring the final volume to 50 μL. Split the mixture into two 25 μL reactions and incubate as follows: 30 s at 98 °C, 15 cycles each of which consists of 15 s at 95 °C, 30 s at 60 °C, and 2 min at 72 °C, and a final extension at 72 °C for 10 min.

10. Pool the PCR reactions and purify the cDNA using the QIAquick PCR Purification Kit, following the manufacturer's

Fig. 2 Illustration of the Y-adapter design. The respective adapters used for *Nla III* and *Tai I*-digested cDNAs are shown on the *left* and *right*, respectively. Primer designations correspond to those in Table 1. The respective overhangs designed to hybridize to the digested cDNA samples are *underlined*

protocol to the letter. Elute the cDNA in 50 μL of the included elution buffer (EB).

11. Split the purified cDNA from **step 10** into two samples of 25 μL each. Mix each sample with 10 μL of the appropriate 10× reaction buffer (provided by the supplier) and bring to a total volume of 100 μL with water. Add 10 U of either of two restriction enzymes (*Nla III* or *Tai I*; *see* **Note 5**) and incubate at the appropriate temperature (37 °C for *Nla III*, 65 °C for *Tai I*) for 1 h. Recover the digested cDNA using the QIAquick PCR Purification Kit, following the manufacturer's protocol to the letter. Elute the cDNA in 50 μL of the included elution buffer (EB).

12. Mix 100 pmol of each oligonucleotide that consists of two strands of a Y adapter (the ADS or ADN series; *see* Fig. 2 and **Note 9**) in a total volume of 100 μL in 10 mM Tris–HCl, pH 7.5 + 0.1 mM Na-EDTA. Bring the mixture to 95 °C and cool slowly (1 °C per minute) until the solution reaches 25 °C.

13. Add 10 pmol of the annealed Y-adapter to 10 μL of the digested cDNA from **step 11**, along with 2 μL of 10× ligase buffer, 400 U of T4 DNA ligase, and water to bring the final volume to 20 μL. Incubate at 4 °C for 12–16 h. Purify the ligated using the QIAquick PCR Purification Kit, following the manufacturer's protocol to the letter. Elute the cDNA in 50 μL of the included elution buffer (EB).

14. Add 1 μL of the purified, ligated cDNA to 5 μL of 5× Phire reaction buffer (included with the enzyme), 2.5 μL of 10× PCR dNTP mix, 10 pmol each of the PE-PCR1 and PE-PCR2 primers (*see* Table 1 and **Note 10**), 0.25 μL of Phire II Hot Start DNA Polymerase, and water to bring the final volume to 25 μL. Execute the reactions as follows: 30 s at 98 °C, between 10 and 20 cycles each of which consists of 15 s at 95 °C, 30 s at 60 °C, and 1 min at 72 °C, and a final extension at 72 °C for 10 min. The cycle number is determined empirically to yield a distribution of DNA ranging between 150 and 700 bp.

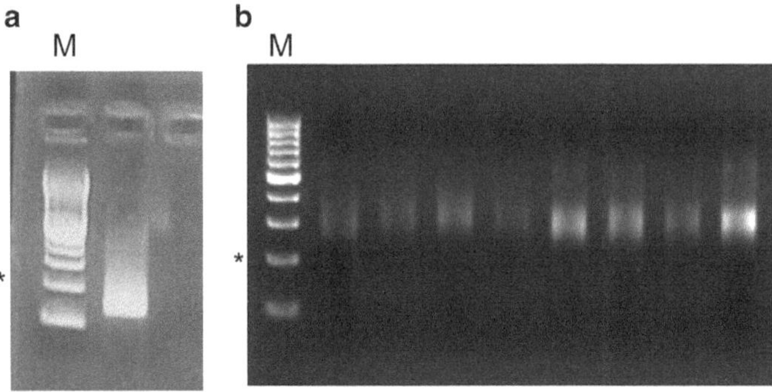

Fig. 3 Examples of samples. All panels are photographs of agarose gels that have been stained with ethidium bromide. (**a**) Tags prepared using Method 1. Lane M—100 bp DNA ladder. The 200 bp band is noted with "*". (**b**) Tags prepared using Method 3. Lane M—100 bp DNA ladder. The 200 bp band is noted with "*". The other lanes represent different independent preparations derived from biological replicates

15. Purify the PCR products using the QIAquick PCR Purification Kit, with a final elution volume of 50 µL (*see* **Note 11**). Assess the resulting preparations using a Bioanalyzer and by agarose gel electrophoresis (Fig. 3a).

3.3.2 Method 2

In Method 2, the ends of cDNA tags that correspond to the 5′ ends of the respective mRNAs are generated by fragmentation of the RNA sample (*see* **Note 12**).

1. Dilute total RNA (between 5 and 50 µg; *see* Subheading 3.2) to 25 µL with RNase-free water. Add 2.8 µL RNA fragmentation buffer and incubate for 8 min at 70 °C (*see* **Note 13**).

2. Add 3 µL RNA fragmentation stop solution and incubate at room temperature for 5 min.

3. Add 105 µL RLT to the reaction, mix, and then add 70 µL of 100 % ethanol (*see* **Note 4**). Mix and apply to a Qiagen RNeasy column (or an equivalent column from other providers). Centrifuge the column in a microcentrifuge at top speed for 30–90 s, discard the flow-through.

4. Apply 500 µL RPE to the column and the sample centrifuged as above. Discard the flow-through and spin the (empty) column at top speed for 2 min.

5. Move the column to a fresh 1.5 mL tube, add 30 µL RNase-free water, and spin at top speed for 30–60 s.

6. Isolate the poly(A)-enriched RNA as described in Subheading 3.3.1, **steps 1–4**; for this, use the entire sample of fragmented RNA. Elute the polyadenylated RNA in 10 µL of 10 mM Tris–HCl, pH 7.5.

7. Mix 9 μL of the polyadenylated fragmented RNA with 1 μL (containing 100 pmol) of the SWITCH2 oligonucleotide (Table 1; *see* **Note 14**) and 1 μL (100 pmol) of one of the PE-RT3 series primers (Table 1; *see* **Note 15**). Heat the mixture to 65 °C for 5 min, chill on ice for 2 min, and repeat.

8. Add the following (while cold): 5 μL SMARTScribe™ 5× first strand buffer, 2.5 μL 10× dNTPs for reverse transcription, 1 μL 100 mM DTT, and 0.5 μL (20 U) RNase Inhibitor. Bring the mixture to 42 °C and add 100 U SMARTScribe™ Reverse Transcriptase. Incubate for 2 h at 42 °C.

9. Inactivate the reverse transcriptase at 70 °C for 5 min, then bring the reaction mixture to 37 °C. Add 1 μL RNase H + 1 μL RNaseA/T1 (*see* **Note 8**) and incubate 37 °C for 30 min.

10. Purify the cDNA using a QIAquick PCR Purification Kit as described in preceding sections. Elute the final cDNA sample into 50 μL.

11. Add 1 μL of the purified cDNA to 5 μL of 5× Phire reaction buffer, 2.5 μL of 10× PCR dNTP mix, 10 pmol each of the PE-PCR1 and PE-PCR2 primers (*see* Table 1), 0.25 μL of Phire II Hot Start DNA Polymerase, and water to bring the final volume to 25 μL. Execute the reactions as follows: 30 s at 98 °C, between 10 and 15 cycles (*see* **Note 16**) each of which consists of 15 s at 95 °C, 30 s at 60 °C, and 1 min at 72 °C, and a final extension at 72 °C for 10 min. The cycle number is determined empirically to yield a uniform distribution of DNA ranging between 150 and 400 bp.

12. Separate the amplicons on a 1 % agarose gel. Excise the region containing fragments between 250 and 400 bp. Purify the DNA with the QIAquick Gel Extraction Kit, eluting the DNA into 50 μL of the elution buffer provided by the supplier.

13. Assess the final purified PCR amplicons on a Bioanalyzer. This is the sample to be submitted for sequencing.

3.3.3 Method 3

In Method 3, first-strand cDNA produced as described in **steps 1–7** of Subheading 3.3.2 are further purified using a series of enrichments and PCR amplifications.

1. Beginning with cDNA as described in **steps 1–7** of the preceding section, add Agencourt® Ampure®XP beads to the cDNA at an 1.8:1 v/v ratio of beads to cDNA (*see* **Note 17**), all in a 200 μL microcentrifuge tube. Mix and incubate at room temperature for 20 min.

2. Separate the beads using a magnetic stand and discard the supernatant. Wash twice with 70 % ethanol (200 μL per wash) and air-dry at room temperature for 15 min. Resuspend in 25 μL water and incubate for 1 min. Remove the beads by magnetic separation and retain the supernatant.

3. Add 1 μL of the purified cDNA to 5 μL of 5× Phire reaction buffer, 2.5 μL of 10× PCR dNTP mix, 10 pmol each of the PE-PCR1 and PE-PCR2 primers (*see* Table 1), 0.25 μL of Phire II Hot Start DNA Polymerase, and water to bring the final volume to 25 μL. Execute the reactions as follows: 30 s at 98 °C, between 10 and 15 cycles each of which consists of 15 s at 95 °C, 30 s at 60 °C, and 1 min at 72 °C, and a final extension at 72 °C for 10 min. The cycle number is determined empirically to yield a uniform distribution of DNA ranging between 150 and 400 bp. Purify amplicons with the QIAquick PCR Purification Kit as described above; the final elution volume is 50 μL.

4. Mix the entire cDNA sample (50 μL) with 20 μL of 5× S1 nuclease reaction buffer (supplied with the enzyme), 30 μL of water, and 100 U of S1 nuclease (*see* **Note 18**). Incubate at 37 °C for 60 min.

5. Separate the S1-digested amplicons on a 1.5 % (w/v) agarose gel run at a very low voltage (between 25 and 50 V) for 3–4 h. Excise the region of the gel containing the desired size range (250–400 bp) of amplicons and purified using the QIAquick Gel Extraction Kit exactly as instructed by the supplier; elute the gel-purified fragments in a final volume of 50 μL.

6. Add 1 μL of the gel-purified PCR product to 5 μL of 5× Phire reaction buffer, 2.5 μL of 10× PCR dNTP mix, 10 pmol each of the PE-PCR1 and PE-PCR2 primers (*see* Table 1), 0.25 μL of Phire II Hot Start DNA Polymerase, and water to bring the final volume to 25 μL. Execute the reactions as follows: 30 s at 98 °C, between 10 and 15 cycles each of which consists of 15 s at 95 °C, 30 s at 60 °C, and 1 min at 72 °C, and a final extension at 72 °C for 10 min (*see* **Note 19**).

7. Separate the amplicons on a 1 % agarose gel. Excise the region containing fragments between 250 and 400 bp. Purify the DNA with the QIAquick Gel Extraction Kit, eluting the DNA into 50 μL.

8. Assess the final purified PCR amplicons on a Bioanalyzer. The final prep may also be assessed by agarose gel electrophoresis (Fig. 3b). This is the sample to be submitted for sequencing (*see* **Note 20**).

4 Notes

1. Do not use the Pine Tree Extraction Buffer immediately after autoclaving; at very high temperatures, the solution is gel-like. It becomes clear and useable at room temperature.

2. The isopropanol precipitation step can be left overnight and is thus a good stopping point for the end of the day.

3. When processing multiple seed samples, **steps 1–4** should be performed for one sample before beginning with the next. The samples may be kept at **step 4** while other samples are being processed, such that all samples may be centrifuged (**step 5**) at one time.

4. This is twice the RLT/sample volume ratio that is recommended by the manufacturer. Our experience is that this higher ratio enhances recovery of RNA, especially if columns other than those provided by Qiagen (such as that may be obtained from numerous lab supply vendors) are used.

5. In Method 1, full-length cDNAs are converted to short forms using restriction enzymes that recognize four-base sequences, and that leave a 4 nt overhang. Enzymes other than those chosen here may be used at the user's discretion. Each cDNA preparation is split into two and digested with one of two enzymes so as to reduce the possible loss of tags whose 3′ ends fall within 1–25 bp of the respective restriction enzyme recognition site. Method 1 is the best of the three methods described here for isolating poly(A) tags derived from very short RNAs, since no size-selection is utilized.

6. The supernatant of the poly(A) enrichment contains unadenylated RNAs, and may be of use in other studies. Thus, it may be recovered using the Qiagen RNA clean-up kit and stored at −80 °C.

7. The SWITCH1.1 primer possesses tandem *Nla III* and *Tai I* recognition sites. Inclusion of these sites is intended to promote the recovery of cDNAs that lack these two sites. The SWITCH1.1 primer also possesses a 3′ oligo dG tract that is intended to facilitate the strand-switching activity of Moloney Murine Leukemia Virus-derived reverse transcriptases such as used here [22]. The RT primers possess randomized positions and a bar code that serves to distinguish different samples when sequenced on the same lane (multiplexing). The RT primers also possess a sequence that renders the tags compatible with a selected sequencing technology; in this report, the sequences are compatible with Illumina HiSeq and MiSeq platforms, but these sequences may be replaced with alternatives based on the respective technology. Some RT primers also have a 5′-biotin group; this provides an option (not described here) for purification and/or removal of first strand cDNAs using streptavidin affinity purification.

8. This is a 20–40 fold excess of the RNAse A/T1 mixture over the capacity of the RNAse inhibitor included in these reactions. Also, the RNAse inhibitor does not work on RNAse H.

9. The so-called Y-adapter is designed to possess a partial double-stranded nature, to anneal to the overhangs left by either *Nla III* or *Tai I*, and to attach an Illumina-compatible sequence to the corresponding end of the cDNA after ligation.

10. The PE-PCR1 and PE-PCR2 primers attach the complete Illumina-compatible sequence to the ends of the cDNA tags. PE-PCR1 corresponds to the end that is sequenced first when choosing paired-end or mate-end sequencing, and should be attached to the appropriate end of tags that will be sequenced from only one direction. If desired, these sequences may be replaced with others suited for the sequencing platform of choice.

11. The final PCR products will have a very disperse range of sizes. This range can be accommodated by the Illumina sequencing technology, based on the consistency of results obtained with Method 1 with those obtained using Methods 2 and 3 [21].

12. In Methods 2 and 3, the ends of the cDNA tags that correspond to the 5′ ends of cDNAs are generated by random fragmentation of RNA. These methods are used to avoid losses of poly(A) sites that may lie close to the sites for restriction enzymes used in Method 1. Method 2 is used with larger quantities (>25 µg) of total RNA, while Method 3 is used for small quantities of total RNA.

13. The fragmentation conditions described here usually yield cDNA tags of sizes less than 500 bp. However, there may be variability in this step, and it is wise to empirically test slightly different fragmentation times and temperatures.

14. The SWITCH2 primers consist of a segment that is compatible with Illumina sequencing, a random sequence, a bar code, and a 3′ GGG. The GGG is intended to promote strand-switching by the SMARTScribe reverse transcriptase.

15. The PE-RT3 series consists of a segment that is compatible with Illumina sequencing, a random dinucleotide, a three-base bar code, a stretch of 18T's, and a 3′-terminal VN dinucleotide; the T18VN serves as an anchored oligo-dT that should anneal preferentially to the poly(A)-mRNA junction.

16. Occasionally, and especially when producing tags from low starting quantities of RNA, sizeable quantities of characteristic artifacts are produced; these artifacts often overwhelm the authentic tags and reduced the overall yield of usable sequence. We have also observed that a number of PCR cycle greater than 15 leads to greater quantities of artifacts. These artifacts are primarily due to dimers formed from the RT and SMART primers during reverse transcription.

17. This volume of AMPure beads to cDNA eliminates DNA molecules shorter than ca. 200 bp in size.

18. Treatment with S1 Nuclease removes single stranded over-hangs of DNA fragments and helps to cleave of dsDNA at single-stranded regions caused by nicks, gaps, mismatches, or hairpin loops. This treatment reduces the abundance of artifacts during the preparation of PATs from low starting quantities of RNA, and thus permits more extensive PCR amplification than would otherwise be possible.

19. The cycle number is determined empirically to yield a uniform distribution of DNA ranging between 250 and 400 bp, with little or no DNA above larger or smaller than these limits. Also, to increase the final yield, several PCR reactions may be pooled into one or two lanes on the final agarose gel.

20. The samples prepared as described in Subheadings 3.3.1, 3.3.2, and 3.3.3 are all ready for submission to sequencing centers; no further preparation or clean-up is needed. Samples prepared using the protocols described here behave similarly in terms of the identification and quantitation of poly(A) sites and mRNA expression levels [21].

Acknowledgements

This work was supported by US National Science Foundation (IOS-0817818) and the University of Kentucky Executive Vice President for Research. Liuyin Ma was a recipient of a scholarship from the China Scholarship Council, and Pratap Kumar Pati was supported by a Fulbright Nehru Senior Research Fellowship.

References

1. Lin Y, Li Z, Ozsolak F, Kim SW, Arango-Argoty G, Liu TT, Tenenbaum SA, Bailey T, Monaghan AP, Milos PM, John B (2012) An in-depth map of polyadenylation sites in cancer. Nucleic Acids Res 40(17):8460–8471. doi:10.1093/nar/gks637

2. Mayr C, Bartel DP (2009) Widespread shortening of 3′UTRs by alternative cleavage and polyadenylation activates oncogenes in cancer cells. Cell 138(4):673–684. doi:10.1016/j.cell.2009.06.016

3. Li Y, Sun Y, Fu Y, Li M, Huang G, Zhang C, Liang J, Huang S, Shen G, Yuan S, Chen L, Chen S, Xu A (2012) Dynamic landscape of tandem 3′ UTRs during zebrafish development. Genome Res 22(10):1899–1906. doi:10.1101/gr.128488.111

4. Smibert P, Miura P, Westholm JO, Shenker S, May G, Duff MO, Zhang D, Eads BD, Carlson J, Brown JB, Eisman RC, Andrews J, Kaufman T, Cherbas P, Celniker SE, Graveley BR, Lai EC (2012) Global patterns of tissue-specific alternative polyadenylation in Drosophila. Cell Rep 1(3):277–289. doi:10.1016/j.celrep.2012.01.001

5. Miura P, Shenker S, Andreu-Agullo C, Westholm JO, Lai EC (2013) Widespread and extensive lengthening of 3′ UTRs in the mammalian brain. Genome Res 23:812. doi:10.1101/gr.146886.112

6. Jan CH, Friedman RC, Ruby JG, Bartel DP (2011) Formation, regulation and evolution of Caenorhabditis elegans 3′UTRs. Nature 469(7328):97–101. doi:10.1038/nature09616

7. Shepard PJ, Choi EA, Lu J, Flanagan LA, Hertel KJ, Shi Y (2011) Complex and dynamic landscape of RNA polyadenylation revealed by PAS-Seq. RNA 17(4):761–772. doi:10.1261/rna.2581711

8. Yoon OK, Hsu TY, Im JH, Brem RB (2012) Genetics and regulatory impact of alternative polyadenylation in human B-lymphoblastoid

cells. PLoS Genet 8(8):e1002882. doi:10.1371/journal.pgen.1002882

9. Hornyik C, Duc C, Rataj K, Terzi LC, Simpson GG (2010) Alternative polyadenylation of antisense RNAs and flowering time control. Biochem Soc Trans 38(4):1077–1081. doi:10.1042/BST0381077

10. Hornyik C, Terzi LC, Simpson GG (2010) The spen family protein FPA controls alternative cleavage and polyadenylation of RNA. Dev Cell 18(2):203–213. doi:10.1016/j.devcel.2009.12.009

11. Liu F, Marquardt S, Lister C, Swiezewski S, Dean C (2010) Targeted 3′ processing of antisense transcripts triggers Arabidopsis FLC chromatin silencing. Science 327(5961):94–97. doi:10.1126/science.1180278

12. Simpson GG, Dijkwel PP, Quesada V, Henderson I, Dean C (2003) FY is an RNA 3′ end-processing factor that interacts with FCA to control the Arabidopsis floral transition. Cell 113(6):777–787

13. Terzi LC, Simpson GG (2008) Regulation of flowering time by RNA processing. Curr Top Microbiol Immunol 326:201–218

14. Thomas PE, Wu X, Liu M, Gaffney B, Ji G, Li QQ, Hunt AG (2012) Genome-wide control of polyadenylation site choice by CPSF30 in Arabidopsis. Plant Cell 24(11):4376–4388. doi:10.1105/tpc.112.096107

15. Zhang J, Addepalli B, Yun KY, Hunt AG, Xu R, Rao S, Li QQ, Falcone DL (2008) A polyadenylation factor subunit implicated in regulating oxidative signaling in Arabidopsis thaliana. PLoS One 3(6):e2410. doi:10.1371/journal.pone.0002410

16. Hunt AG, Xing D, Li QQ (2012) Plant polyadenylation factors: conservation and variety in the polyadenylation complex in plants. BMC Genomics 13:641. doi:10.1186/1471-2164-13-641

17. Mueller AA, Cheung TH, Rando TA (2013) All's well that ends well: alternative polyadenylation and its implications for stem cell biology. Curr Opin Cell Biol 25(2):222–232. doi:10.1016/j.ceb.2012.12.008

18. Sherstnev A, Duc C, Cole C, Zacharaki V, Hornyik C, Ozsolak F, Milos PM, Barton GJ, Simpson GG (2012) Direct sequencing of Arabidopsis thaliana RNA reveals patterns of cleavage and polyadenylation. Nat Struct Mol Biol 19(8):845–852. doi:10.1038/nsmb.2345

19. Shi Y (2012) Alternative polyadenylation: new insights from global analyses. RNA 18(12):2105–2117. doi:10.1261/rna.035899.112

20. Wu X, Liu M, Downie B, Liang C, Ji G, Li QQ, Hunt AG (2011) Genome-wide landscape of polyadenylation in Arabidopsis provides evidence for extensive alternative polyadenylation. Proc Natl Acad Sci U S A 108(30):12533–12538. doi:10.1073/pnas.1019732108

21. Ma L, Pati PK, Liu M, Li QQ, Hunt AG (2013) High throughput characterizations of poly(A) site choice in plants. Methods 67:74. doi:10.1016/j.ymeth.2013.06.037

22. Zhu YY, Machleder EM, Chenchik A, Li R, Siebert PD (2001) Reverse transcriptase template switching: a SMART approach for full-length cDNA library construction. Biotechniques 30(4):892–897

Chapter 15

DNA/RNA Hybrid Primer Mediated Poly(A) Tag Library Construction for Illumina Sequencing

Man Liu, Xiaohui Wu, and Qingshun Quinn Li

Abstract

Alternation polyadenylation is widespread in eukaryotes, and has demonstrated roles in gene expression regulation. Owing to deep DNA sequencing technologies, global analyses of alternation polyadenylation and their functions have become possible. We present a method to generate poly(A) tags libraries for high-throughput sequencing (PAT-seq). This protocol targets the junction of the 3′-UTR and poly(A) tail of a transcript so it can be positively identified as a poly(A) site. Upon Zinc-mediated limited digestion of total RNA, RNA fragments with poly(A) tail are then isolated and 5′-end repaired. A DNA/RNA hybrid adaptor is ligated to the 5′ end as an anchor. Then the library is generated by reverse transcription with oligo(dT)-adapter followed by PCR amplification. Such a custom poly(A) tags library can be generated from any source poly(A) containing RNA and good for both single- or paired-end sequencing in any Illumina sequencing platforms. This new method has been applied to investigate mRNA polyadenylation in Arabidopsis.

Key words Polyadenylation, Illumina sequencing, Arabidopsis, PAT-seq, DNA/RNA hybrid primer

1 Introduction

Increasing evidences implicate the important role of alternative polyadenylation (APA) of mRNA in the regulation of gene expression in eukaryotes [1–3], impacting nervous system development, immune responses, cancer and embryo development in animals, and flowering time and stress responses in plants. Genome-wide studies have showed extensively alternative polyadenylation in many organisms. It has been reported that more than half of the human genes and the mouse genes can have alternative polyadenylation sites [4, 5], and about 2,000 genes have three or more APA isoforms in the human and mouse cells [5]. In Arabidopsis, over 70 % genes has more than one poly(A) site, and a large number of APA are in introns and coding sequences [6], which would lead to change the coding sequence and thus providing the functional diversity. More and more results showed that APA is

Arthur G. Hunt and Qingshun Quinn Li (eds.), *Polyadenylation in Plants: Methods and Protocols*, Methods in Molecular Biology, vol. 1255, DOI 10.1007/978-1-4939-2175-1_15, © Springer Science+Business Media New York 2015

involved in many aspects of gene expression regulation than previously anticipated. The APA of mRNA can generate different protein isoforms with diverse functions, and transcripts with distinct 3′UTR that is significant to mRNA stability and translation efficiency. Recently, a surprisingly large fraction of polyadenylation sites in antisense transcripts were found in plants [6]. How these regulate gene expression is unknown. Characterizing APA globally would help us to understand posttranscriptional gene regulation.

In the past, the analysis of plant APA dataset has been hindered by the lack of reliable methods to collect and quantify APA. Previous studies have been using the results based on ESTs [4, 7], or microarray analyses [8], which both were not specially designed for the APA, and had the major limitations. For example, the EST methods are low throughput, and the microarray cannot directly target to the exactly ploy(A) sites. With the advantage of high-throughput sequencing platforms, a number of methods have been developed to identify the mRNA-poly(A) junction [5, 6, 9]. Jan et al. used a method called poly(A)-position profiling by sequencing (Fig. 1) to identify of polyadenylated RNA termini in *Caenorhabditis elegans* [9]. In this method, the biotinylated adaptor was first ligated to the 3′ end of oligo(dT) selected RNA. The ligation products were then partially digested by T1 nuclease, the polyadenylated 3′-ends purified. After reverse transcription with dTTP as the only nucleotide,

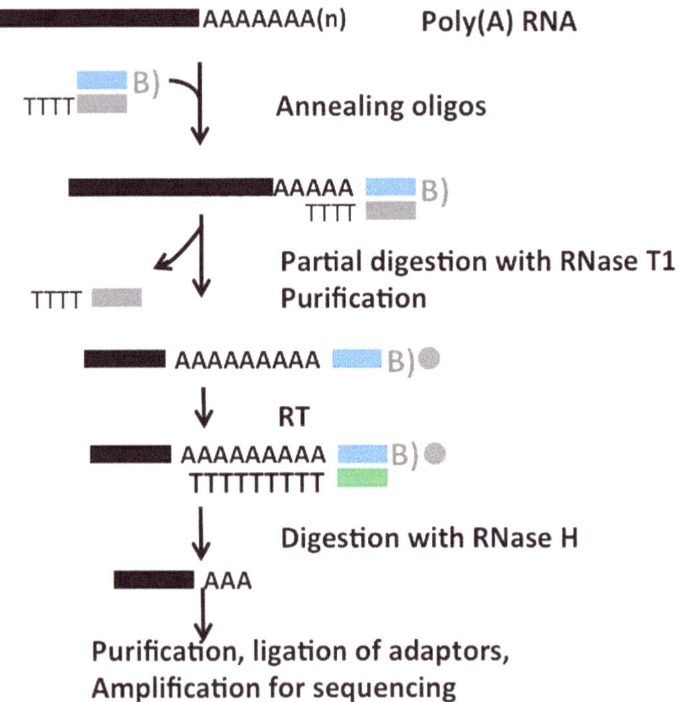

Fig. 1 Illustration of the method of 3P-seq (based on Jan et al. [9])

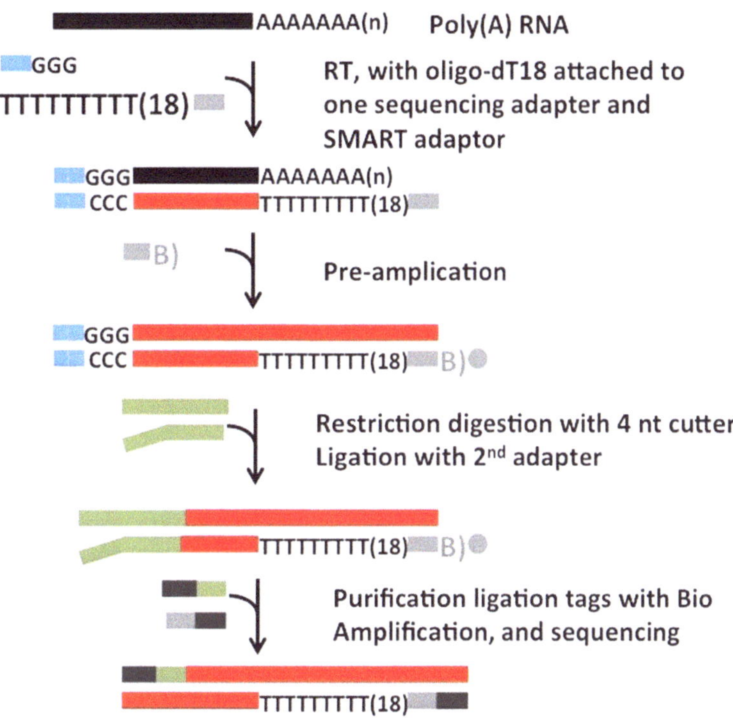

Fig. 2 Illustration of the method of PAT-Seq in Wu et al. [6]

the RNase H digestion freed most part of the poly(A) region, and the remaining tag were purified, ligated to primers for high-throughput sequencing. At the same time, a new deep-sequencing protocol were used for genome-wide analysis of polyadenylation in Arabidopsis [6], and a Poly(A) Site Sequencing (PAS-Seq) protocol were developed to reveal the APA in human and mouse, and in the stem cell differentiation [5]. These two methods share one common point. The MMLV reverse transcriptase has terminal a transferase activity, which can adds three cytosine deoxyribonucleotides to the 3′ end of the first strand of cDNA. For adding one of the deep sequencing adaptors, a hybrid oligonucleotide adaptor with three guanine ribonucleotides at the 5′ end annealed to the reverse transcription products. For the deep sequencing protocol by Wu et al. (Fig. 2), after the second strand synthesis, restriction enzyme recognizing four nucleotides was used to digest cDNA into smaller pieces. The digested cDNA was then ligated to a Y-adapter which has primers with partial sequence complementarity and mismatching at the appropriate ends. In PAS-seq method, Poly(A) enriched RNAs were firstly fragmented to 6–~200 nucleotide by using a chemical method. The fragmented poly(A) RNA as the template performed the reverse transcription (Fig. 3).

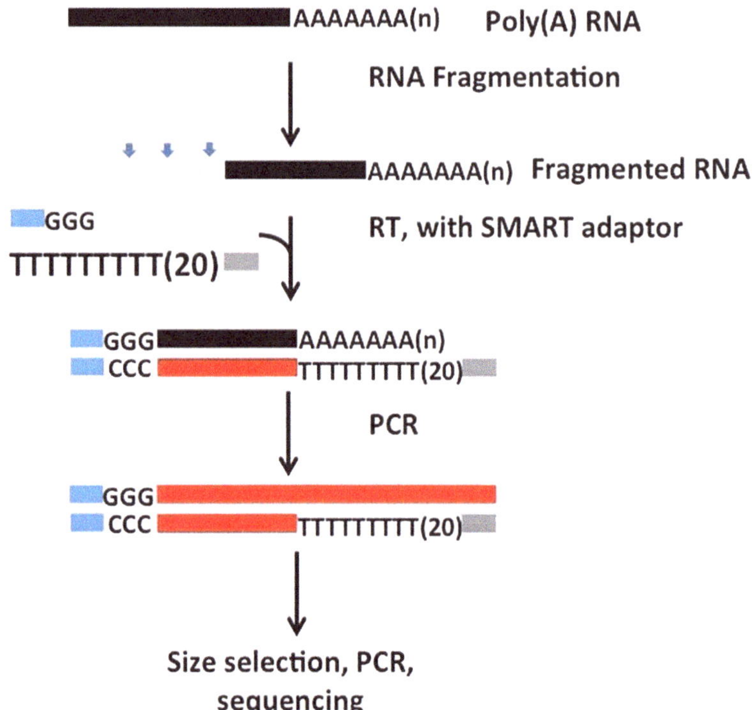

Fig. 3 The schematic of PAS-Seq (based on Shepard et al. [5])

Here, we have developed a new method, Poly(A) tag sequencing method (PAT-seq) for the deep and global analyses of 3′ ends of mRNA. It has been used to study mRNA polyadenylation in the Arabidopsis root, and initial analysis of the results indicated its validity in collecting large-scale poly(A) sites.

As outlined in Fig. 4, this procedure starts with a total RNA fragmentation to the desired size range from 200 to 400 bp (*see* **Note 1**). Next, the fragmented mRNAs with poly(A) tail are purified by oligo (dT) magnetic beads. The DNA/RNA adaptor primer with part of Illumina PE2 primer sequence are ligated to the 5′ end of the end-repaired fragmented RNA, and double-stranded cDNA is then synthesized with a bar-coded Illumina oligonucleotide. Finally, the cDNAs are amplified and size selected to produce a library for deep-sequencing. This library is best suited for single-end Illumina sequencing. However, with slight modification, it can be easily adopted for paired-end sequencing in Illumina or other platforms.

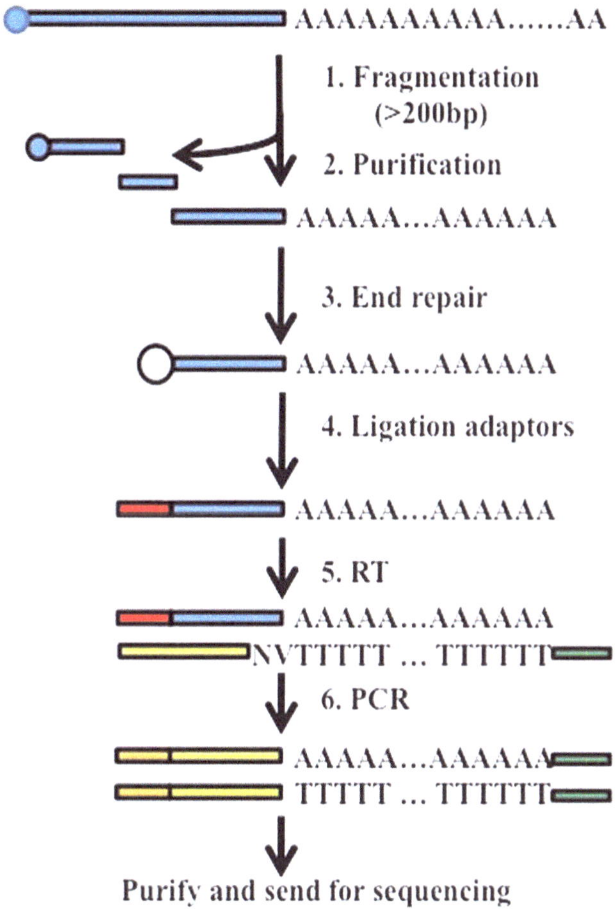

Fig. 4 Schematic of the PAT-seq protocol. The detail is described in the text. *Blue*, *red*, and *yellow boxes* represent RNA, adaptors, and cDNA; *green* and *orange boxes* are primers. V represents A/C/G, N represents A/T/C/G

2 Materials

1. Total RNA isolated from Arabidopsis roots.
2. Solutions:

   10× Fragmentation Buffer: 100 mM $ZnCl_2$ in 100 mM Tris–HCl pH 7.0.

   Binding Buffer: 20 mM Tris–HCl pH 7.5, 1.0 M LiCl, and 2 mM EDTA.

   Washing Buffer B: 10 mM Tris–HCl PH 7.5, 0.15 M LiCl, 1 mM EDTA.

   Stop Solution: 0.5 M EDTA, pH 8.0.
3. RNA purification Kit (e.g., Qiagen RNA RNeasy Kit).

Table 1
The sequences of adaptors and primers

Name	Sequence (from 5′–3′)
DNA/RNA Hybrid Adaptor	(OH)CGGTCTCGGCATTCCTGCTGAArCrCrGrCrUrCrUrUrCrCrGrArUrCrU[a]
RT-SE3c	ACACTCTTTCCCTACACGACGCTCTTCCGATCTNN<u>AAC</u>TTTTTTTTTTTTTTTTTTVN (*see* **Note 5**)[b]
RT-SE3f	ACACTCTTTCCCTACACGACGCTCTTCCGATCTNN<u>AGA</u>TTTTTTTTTTTTTTTTTVN
RT-SE3k	ACACTCTTTCCCTACACGACGCTCTTCCGATCTNN<u>GGA</u>TTTTTTTTTTTTTTTTTVN
RT-SE3n	ACACTCTTTCCCTACACGACGCTCTTCCGATCTNN<u>GAG</u>TTTTTTTTTTTTTTTTTVN
PE-PCR1	AATGATACGGCGACCACCGAGATCTACACTCTTTCCCT*ACACGACGCTCTTCCGATCT* (*see* **Note 6**)
PE-PCR2	CAAGCAGAAGACGGCATACGAGAT*CGGTCTCGGCATTCCTGCTGAACCGCTCTTCCGATCT*

[a](OH) at the 5′-end of the primer prevents it from ligation; r denotes a ribonucleotide
[b]V = A/C/G; N = A/T/C/G; underlined nucleotides are barcodes

4. Oligo(dT)$_{25}$ Magnetic Beads.

5. 10× T4 polynucleotide kinase.

6. T4 RNA ligase 1.

7. Reverse transcriptase (e.g., SuperScript III).

8. RNase H.

9. DNA gel purification kit (e.g., QIAquick Gel extraction kit).

10. Primers (*see* Table 1).

3   Methods

3.1   RNA Fragmentation

1. Preheat a thermal cycler to 70 °C.

2. Prepare the following reaction mix in a PCR tube:
   - 10× Fragmentation Buffer (2 μl).
   - RNA (2 μg total RNA).
   - Water.

   The total volume should be 20 μl.

3. Incubate the tube in the preheated thermal cycler for 5 min.

4. Add 2 μl of Fragmentation Stop Solution and incubate the sample at room temperature (RT) for 5 min.

5. Follow the instructions in the Qiagen RNA RNeasy Kit to purify fragmented RNA, and elute in 50 μl DEPC treated H₂O.

3.2 Poly(A) Enrichment

1. Heat the fragmented RNA at 65 °C for 5 min to disrupt the secondary structures, and place the tube on ice.

2. Meanwhile, resuspend Oligo(dT)$_{25}$ beads by agitating at RT for 30 min, and aliquot 20 μl of oligo(dT) beads (*see* **Note 2**).

3. Wash the beads twice with 100 μl of Binding Buffer, and remove the supernatant.

4. Resuspend the beads in 50 μl of Binding Buffer, and add the 50 μl of the above heat fragmented RNA, rotate the tube at RT for 10 min, and remove the supernatant.

5. Wash the beads twice with 100 μl of Washing Buffer B.

6. Add 22 μl of DEPC treated H₂O and heat the beads in 80 °C water bath for 2 min to elute mRNA.

3.3 5′-End Repair

1. Prepare a heat block at 37 °C.

2. Mix the following reagents in a 1.5 ml RNase-free tube to total 50 μl volumes.
   - Poly(A) enriched RNA (20 μl).
   - 10× T4 polynucleotide kinase reaction Buffer (5 μl).
   - 100 mM ATP (0.5 μl).
   - T4 polynucleotide kinase (2 μl).
   - DEPC treated H₂O (22.5 μl).

3. Incubate at 37 °C for 30 min and put on ice immediately.

4. Purify products by using a RNA isolation column (e.g., Qiagen RNeasy Kit, *see* Subheading 3.2), and elute in 40 μl DEPC treated H₂O (*see* **Note 3**).

3.4 Ligate the DNA/ RNA Hybrid Adapter (See Note 4)

1. Mix the following reagents in a 1.5 ml RNase-free tube to a total of 100 μl:
   - End repaired RNA above.
   - 10× T4 RNA ligase Buffer (10 μl).
   - 100 mM ATP (1 μl).
   - 100 mM Hexammine cobalt chloride (1 μl).
   - 50 % PEG 8000 (50 μl).
   - 100 μM DNA/RNA adapter (2 μl) (*see* Table 1).
   - T4 RNA ligase (1 μl).

2. Incubate at 22 °C for 16 h.

3. Purify the ligation products by a RNA isolation column (e.g., Qiagen RNeasy Kit), elute in 25 μl DEPC treated H₂O.

**3.5 Synthesize
the First Strand cDNA**

1. Mix the following reagents in a 200 μl PCR tube:
   - 5′ Ligated RNA from Subheading 3.4 above (11 μl).
   - 100 μM RT-PE3 primer (1 μl) (*see* Table 1).
   - 10 mM dNTP (1 μl).

2. Incubate the tube at 65 °C for 5 min and chill the tube on ice immediately. Add the following reagents to the PCR tube:
   - 5× First Strand Buffer (4 μl).
   - 100 mM DTT (1 μl).
   - RNase inhibitor (e.g., RNase Out) (1 μl).
   - Reverse transcriptase (e.g., SuperScript III) (1 μl).

   Mix well, and the total volume should be 20 μl.

3. Put the mixture in a PCR thermal cycler with the program of 48 °C for 1 h, 70 °C for 15 min, and hold at 4 °C.

4. Add 1 μl RNase H and incubate at 37 °C for 50 min.

**3.6 Amplify
the cDNA Templates**

1. Prepare the PCR mixture for 20 μl × 3 tube:
   - 2× Fusion PCR Master Mix (10 μl).
   - 10 μM PE-PCR1 primer (1 μl) (*see* Table 1).
   - 10 μM PE-PCR2 primer (1 μl) (*see* Table 1).
   - Water (6 μl).
   - cDNA template from Subheading 3.5 above (2 μl).

   Mix well.

2. The following PCR program is used to amplify the products:
   (a) 98 °C for 30 s.
   (b) 15 cycles of:

       98 °C for 10 s.
       60 °C for 30 s.
       72 °C for 60 s.

   (c) 72 °C for 10 min.
   (d) Hold at 4 °C.

**3.7 Purify the cDNA
Library**

1. Load 2 μl 100 bp DNA ladder, 50 μl the cDNA library mixed with 5 μl 10× DNA Loading Dye, 50 μl PCR control products amplified by pE-PCR2 single primer to 2 % agarose gel.

2. Cut the region of the gel between 300 and 500 bp.

3. Use DNA gel extraction kit (e.g., QIAquick Gel extraction kit) to purify the DNA sample, and elute in 20 μl water.

4. The purified product is ready for quality control analysis by qPCR, cloning-Sanger sequencing, and/or Agilent 2100 DNA-chip. After that, the sample is ready for Illumina sequencing.

4   Notes

1. Ten-day old roots of *Arabidopsis thaliana* (ecotype Colombia) vertically grown in MS plate are used to construct the PAT-seq library. However, if the materials changed, the temperature and the time for the RNA fragmentation should also be tested to achieve ideal results. Since the average length of 3′-UTR in Arabidopsis is about 225 nucleotides [7], to get more meaningful RNA fragment with both poly(A) tail and RNA coding information, the aimed size of fragmented RNA is about 200 nucleotides.

2. It is a good practice to aliquot the Oligo(dT)$_{25}$ beads into small, working fractions and store in 4 °C refrigerator when it first arrives. This will avoid repeated warming and agitating at RT, which may potentially reduce the functionality of oligo (dT) and the beads.

3. The purification of modified poly(A) RNA fragment should be done within 1 h after the T4 polynucleotide kinase treatment.

4. DNA/RNA hybrid primer, instead of all RNA oligo primer, is used here for more stable and economical purposes. All RNA primer would cost more to make and less stable. Also, due to DNA nucleotides are on the 5′-end, ligation of the hybrid primer to 3′-end of RNA (or primer self-ligation) is avoided by RNA ligase.

5. Bar-coding, as it is implemented (underlined in Table 1) in the RT primers, is used to allow pooling different samples together in an Illumina sequencing lane. The two random nucleotides before the bar-codes are to increase heterogeneity in order to allow cluster recognition by Illumina sequencing protocols. Longer barcoding or longer random nucleotides may be desirable if the length of sequencing is not a limiting factor.

6. The italicized sequences in PE-PCR1 and 2 match the RT primers and DNA/RNA hybrid adapter, respectively. This allow for PCR purification and the attachment of Illumina sequencing primers.

7. We obtained a total of 35,030,150 valid poly(A) sites in Arabidopsis root. Because of microheterogeneity in plant mRNA 3′-ends, the poly(A) sites located within 24 nucleotides were clustered. Thus, these poly(A) sites can be classified into 40,227 Poly(A) Cluster (PAC). The distribution of PAC is stated in Table 2.

Table 2
The distribution of poly(A) clusters in the root of Arabidopsis ecotype Columbia

Annotated genomic region	Number of PAC	Percentage of PAC
3′UTR	27,848	69.23
Intergenic	3,948	9.81
Promoter	2,436	6.06
CDS	2,118	5.27
Intron	1,152	2.86
5′UTR	205	0.51
Exon	424	1.05
Pseudogenic_exon	87	0.22
Ambiguous	2,009	4.99
Total	40,227	100

PAC poly(A) site cluster, *CDS* coding sequence, *UTR* untranslated region

Acknowledgement

We thank other lab members for helpful discussion and testing of the protocol. This work was supported by US National Science Foundation (grant nos. IOS–0817829 and IOS-1353354 to QQL), and a grant from Ohio Plant Biotech Consortium.

References

1. Elkon R, Ugalde AP, Agami R (2013) Alternative cleavage and polyadenylation: extent, regulation and function. Nat Rev Genet 14(7):496–506
2. Xing D, Li QQ (2011) Alternative polyadenylation and gene expression regulation in plants. Wiley Interdiscip Rev RNA 2(3):445–458
3. Shi Y (2012) Alternative polyadenylation: new insights from global analyses. RNA 18(12): 2105–2117
4. Tian B, Hu J, Zhang H, Lutz CS (2005) A large-scale analysis of mRNA polyadenylation of human and mouse genes. Nucleic Acids Res 33(1):201–212
5. Shepard PJ et al (2011) Complex and dynamic landscape of RNA polyadenylation revealed by PAS-Seq. RNA 17(4):761–772
6. Wu X et al (2011) Genome-wide landscape of polyadenylation in Arabidopsis provides evidence for extensive alternative polyadenylation. Proc Natl Acad Sci U S A 108(30):12533–12538
7. Loke JC et al (2005) Compilation of mRNA polyadenylation signals in Arabidopsis revealed a new signal element and potential secondary structures. Plant Physiol 138(3):1457–1468
8. Ji Z, Lee JY, Pan Z, Jiang B, Tian B (2009) Progressive lengthening of 3′ untranslated regions of mRNAs by alternative polyadenylation during mouse embryonic development. Proc Natl Acad Sci U S A 106(17):7028–7033
9. Jan CH, Friedman RC, Ruby JG, Bartel DP (2011) Formation, regulation and evolution of Caenorhabditis elegans 3′UTRs. Nature 469(7328):97–101

Chapter 16

Poly(A) Tag Library Construction from 10 ng Total RNA

Jingyi Cao and Qingshun Quinn Li

Abstract

Alternative polyadenylation has been demonstrated as a tier of gene expression regulation in eukaryotes. However, its role has not been elucidated at the cellular level. Equipped with techniques to isolate single cells by fluorescence-activated cell sorting (FACS) and laser captured micro-dissection, analysis of alternative polyadenylation in specific cell types becomes possible. We present a method to generate poly(A) tags for high-throughput sequencing (PAT-seq) libraries from very low amount of total RNA. This protocol targets the junction of the 3′-UTR and poly(A) tail of transcripts. Ten nanograms of total RNA isolated from the FACS-sorted cells was reverse-transcribed to double stranded cDNA with a anchored oligo $dT_{(18)}$ primer containing maximal T7 promoter sequence. Then, an RNA amplification step using in vitro transcription of T7 RNA polymerase was carried out. Achieved cRNA was fragmented by partial digestion. First strand synthesis was carried out by using a partial adaptor sequence with random 9-nt primer to introduce the adaptor at the 5′ end. An anchored oligo dT primer containing adaptor sequence on 3′ end was introduced through second strand cDNA synthesis. This new method has been applied to investigate polyadenylation using nanogram amount of total RNA from Arabidopsis cells.

Key words Polyadenylation, Illumina sequencing, Arabidopsis, T7 in vitro Transcription, PAT-seq

1 Introduction

As an essential step in mature mRNA formation, polyadenylation is processed coupling with capping and splicing in order to build up the 3′ end properly. Performed as a tier of gene expression regulation, the length and specific sequences in the generated 3′ un-translated region (UTR) have been shown to be responsible for many properties of mRNAs such as specific cytoplasmic localization, nuclear transport, mRNA turnover and translation initiation [1]. With the advances of large-scale sequencing and bioinformatics analysis, significant amount of alternative polyadenylation (APA) has been discovered among different species in genome-wide level [2–5]. In Arabidopsis, around 70 % of genes were shown to possess APA sites [5]. Being correlated to the functions of 3′ end, the alternative cleavage and polyadenylation produced by APA events have been revealed to affect gene expression both qualitatively and quantitatively [6]. Moreover,

Arthur G. Hunt and Qingshun Quinn Li (eds.), *Polyadenylation in Plants: Methods and Protocols*, Methods in Molecular Biology, vol. 1255, DOI 10.1007/978-1-4939-2175-1_16, © Springer Science+Business Media New York 2015

specific transcripts generated by APA are suggested as regulative products to spatial and developmental environments [7], thus any investigation on cell type-APA will be a potential tool to solve the puzzles of alternative RNA processing in multicellular organisms. Recent studies in T cells activation and cancer lines revealed the preferential of APA site usage was also affected upon cell differentiation and developmental stages [6]. Previous studies also suggest tissue-specific APA site usages in human tissues [8] and the lifespan of worms [4], and their APA preferences were revealed to be species-specific [9]. However, limited reports on alternative RNA processing events in a cell type-specific manner in the plant world. Even though the flowering time control in the model plant Arabidopsis has been explained and considered as a best example of APA from last two decades, the alternative 3′ end processing in specific plant cells remained unknown until now.

To analyze the alternative cleavage and polyadenylation in a specific cell-type level, a method incorporates linear amplification following with poly(A) tag (PAT) library need to be established for the minute amount of total RNA extracted from isolated cells. Due to its isothermal incubation, T7 RNA polymerase mediated in vitro transcription has been considered as a linear amplification process. Besides of synthesizing RNA from specific genes, in vitro transcription has also been applied in generation of cDNA library [10]. Here we introduce a PAT library procedure using T7 RNA polymerase transcription as an amplification step. As shown in Fig. 1, total RNA is first reverse transcribed with an oligo d(T) primer containing maximal T7 promoter sequence. Double-stranded cDNA incorporated with T7 promoter sequence is used as template for T7 in vitro transcription. The generated antisense RNA (aRNA) is fragmented to an average size of 200 nt. The fragmented aRNA is reverse transcribed with a random 9-nt primer containing partial of Illumina PE2 adaptor sequence. Partial adaptor containing Illumina PE1 with barcoding is introduced to the 3′ end during extension by Klenow enzyme. The generated double-stranded cDNA library is finally amplified using Illumina PE1 and PE2 adaptors and size selected. Here we used the total RNA isolated from fluorescence-activated cell sorting (FACS) cells to construct a PAT-seq library. This protocol can be applied to Illumina library prep from nanograms of total RNA collected from cell sorting events.

2   Materials

1. Total RNA isolated from fresh Arabidopsis leave root cells (*see* **Note 1**).

2. Reverse Transcriptase (e.g., SuperScript III).

3. *E. coli* DNA polymerase I.

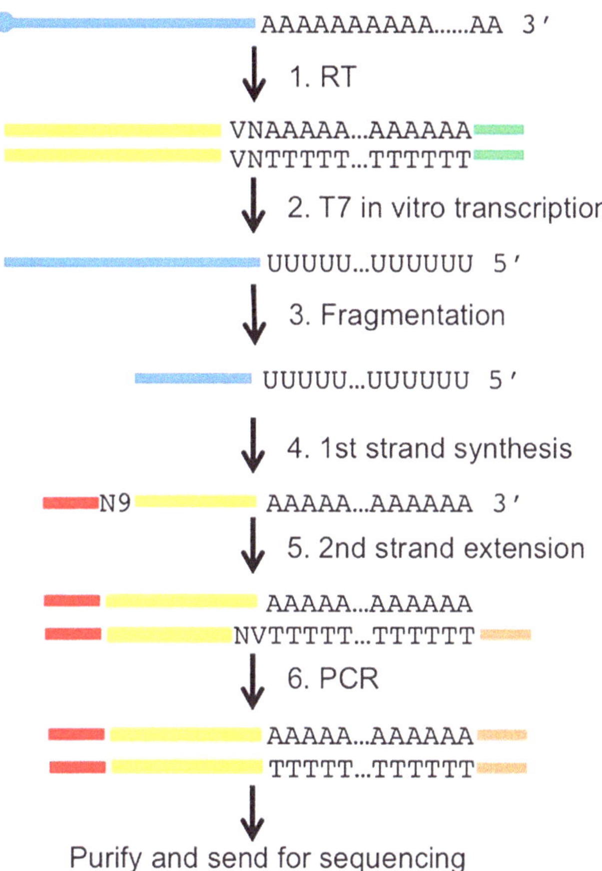

Fig. 1 Schematic of the PAT-seq protocol. The details are given in the text. *Blue* and *yellow boxes* represent RNA, and cDNA; *red*, and *orange boxes* are Illumina adaptors; *green boxes* are primers containing T7 promoter sequence. D represents A, G, or T; V represents A. C or G, N, any nucleotide. The orientation of the RNA/DNA is marked by 5′ or 3′

4. *E. coli* DNA ligase.

5. T4 DNA polymerase.

6. RNase H.

7. PCR Purification Kit (e.g., NucleoSpin Gel and PCR Cleanup kit).

8. T7 in vitro Transcription Kit (e.g., AmpliScribe T7 High Yield Transcription Kit).

9. RNA Purification Kit (e.g., NucleoSpin RNA Cleanup Kit).

10. Klenow Fragment ($3′ \rightarrow 5′$ exo⁻).

11. DNA Gel Purification Kit (e.g., NucleoSpin Gel and PCR Cleanup kit).

12. Primers (*see* Table 1).

Table 1
The sequences of adaptors and primers

Name	Sequence (from 5′–3′)
T7-T(18)	GGCCAGTGAATTGTAATACGACTCACTATAGGGAGGCGGTTTTTTTTTTTTTTTTTT(A/C/G)(A/C/G/T)
PE2-20-9N	CTGAACCGCTCTTCCGATCTNNNNNNNNN
RT-SE3n	ACACTCTTTCCCTACACGACGCTCTTCCGATCTNN<u>GAG</u>TTTTTTTTTTTTTTTTTTT(A/C/G)(A/C/G/T) (*see* **Note 9**)
PE-PCR1	AATGATACGGCGACCACCGAGATCTACACTCTTTCCCT*ACACGACGCTCTTCCGATCT* (*see* **Note 10**)
PE-PCR2	CAAGCAGAAGACGGCATACGAGATCGGTCTCGGCATTCCTGC*TGAACCGCTCTTCCGATCT*

3   Methods

3.1 Synthesis of the First Strand cDNA

1. Mix the following reagents in a 200 μl PCR tube:
   - Total RNA (10 ng).
   - 0.5 μg T7-T18 primer (1 μl) (*see* Table 1; **Note 2**).
   - 10 mM dNTP mix (1 μl).
2. Incubate the tube at 65 °C for 5 min, and chill the tube on ice immediately.
3. Add the following reagents to the PCR tube:
   - 5× First Strand Buffer (2 μl).
   - 100 mM DTT (1 μl).
   - RNase Inhibitor (e.g., RNase OUT) (1 μl).
4. Incubate the tube at 42 °C for 5 min.
5. Add 1 μl of Reverse Transcriptase (e.g., SuperScript III).
6. Mix well, and the total volume should be 10 μl.
7. Put the mixture in a PCR thermal cycler with the program of 42 °C for 1 h, 70 °C for 15 min, and hold at 4 °C.

3.2 Synthesis of the Second Strand cDNA

1. Add the following reagents to the first strand-cDNA above:
   - 10 mM dNTP mix (1 μl).
   - 10× NEBuffer 2 (2 μl) (10 mM Tris–HCl working buffer containing 50 mM NaCl, 10 mM MgCl$_2$, and 1 mM DTT, from New England Biolabs).
   - 1 M (NH$_4$)$_2$SO$_4$ (1.5 μl).
   - *E. coli* DNA polymerase I (1 μl).
   - RNase H (0.5 μl).

2. Mix well, and the total volume should be 20 µl.

3. Incubate the tube at 16 °C for 2 h.

4. Add the following reagents to above reaction:
   - 10× *E. coli* DNA ligase buffer (2 µl).
   - *E. coli* DNA ligase (0.5 µl).

5. Incubate at room temperature for 15 min.

6. Then add the following reagents:
   - 10 mM dNTP mix (1 µl).
   - T4 DNA polymerase (0.5 µl).

7. Incubate at room temperature for 15 min.

8. Follow the instructions in the PCR Purification Kit to purify double stranded cDNA, and elute in 20 µl nuclease-free water.

3.3 T7 In Vitro Transcription

1. Mix the following reagents in a 200 µl PCR tube to a total 20 µl:
   - Purified double stranded cDNA above (10 µl).
   - 10× Reaction Buffer (2 µl) (*see* **Note 3**).
   - 100 mM NTP mix (6 µl) (*see* **Note 4**).
   - 100 mM DTT (1 µl).
   - T7 RNA polymerase (2 µl).
   - RNase Inhibitor (0.5 µl).

2. Incubate at 42 °C for 3.5 h (*see* **Note 5**).

3. Add 1 µl of TURBO DNase (RNase free DNase). Mix well and incubate at 37 °C for 15 min.

4. Follow the instructions in the RNA Cleanup Kit to purify the cRNA, and elute in 30 µl nuclease-free water.

3.4 RNA Fragmentation

1. Preheat a thermal cycler to 94 °C.

2. Prepare the following reaction mix in a 200 µl PCR tube:
   - 5× First Strand Buffer (4 µl).
   - Purified cRNA (8 µl).

   Mix well.

3. Incubate the tube in the preheated thermal cycler for 3 min (*see* **Note 6**).

4. Chill the reaction mix on ice and proceed to next step.

3.5 Synthesis of the First Strand cDNA with T7 Amplified RNA

1. Mix the following reagents in a 200 µl PCR tube:
   - Fragmented cRNA above (Subheading 3.4) (12 µl).
   - 0.5 µg PE2-20-9N primer (1 µl).
   - 10 mM dNTP mix (1 µl).

2. Incubate the tube at 65 °C for 5 min, and chill the tube on ice immediately.

3. Add the following reagents to the PCR tube:
   - 100 mM DTT (1 µl).
   - RNase Inhibitor (e.g., RNase OUT) (0.5 µl).

4. Incubate the tube at 25 °C for 10 min.

5. Add 1 µl of Reverse Transcriptase (e.g., SuperScript III).

6. Mix well, and the total volume should be 20 µl.

7. Put the mixture in a PCR thermal cycler with the program of 42 °C for 1 h, 70 °C for 15 min.

8. Add 1 µl RNase H, incubate at 37 °C for 50 min, then at 65 °C for 20 min.

9. Follow the instructions for PCR Cleanup kit to purify the first-strand cDNA, and elute in 20 µl nuclease-free water.

3.6 Synthesis of the Second Strand cDNA

1. Mix the following reagents in a 200 µl PCR tube:
   - Purified first-strand cDNA above (11 µl).
   - 10 µM RT-SE3n primer (1 µl) (*see* Table 1; **Note 7**).
   - 10 mM dNTP mix (1 µl).

2. Incubate the tube at 98 °C for 3 min, then ramp down to 4 °C at the rate of –0.1 °C/s. Chill the tube on ice immediately.

3. Add the following reagents to the PCR tube:
   - 10× NEBuffer 2 (2.5 µl).
   - 10 mM dNTP mix (1 µl).
   - Klenow (3′–5′ exo⁻) (1 µl).

4. Incubate the tube at 37 °C for 1 h.

5. Follow the instructions of PCR Clean-up kit to purify the first-strand cDNA, and elute in 20 µl nuclease-free water.

3.7 Amplification of the cDNA Library

1. Mix the following reagents in a 200 µl PCR tube:
   - Purified double strand cDNA from above Subheading 3.6 (11 µl).
   - 10 µM PE-PCR1 primer (0.5 µl) (*see* Table 1).
   - 10 µM PE-PCR2 primer (0.5 µl) (*see* Table 1; **Note 8**).
   - 10× PCR Buffer (2 µl).
   - 50 mM MgCl$_2$ (0.8 µl).
   - 10 mM dNTP mix (1 µl).
   - Taq DNA polymerase (1 µl).

2. The following PCR program is used to amplify the products:

(a) 94 °C for 2 min.

(b) 25 cycles of:

94 °C for 15 s.

60 °C for 30 s.

72 °C for 30 s.

(c) 72 °C for 10 min.

(d) Hold at 4 °C.

3.8 Purification and Size Selection of the cDNA Library

1. Clean DNA electrophoresis gel box and replace with new 1× TBE buffer.

2. Load 4 μl 100 bp DNA ladder, 20 μl of the cDNA library mixed with 3 μl 6× DNA loading dye, to ethidium bromide containing 2 % agarose gel.

3. Run the 7 cm gel at 110 V for 20 min.

4. Cut the region of the gel between 200 and 500 bp.

5. Use DNA gel extraction kit (e.g., NucleoSpin Gel and PCR purify kit) to purify the DNA sample, and elute in 20 μl water.

3.9 Quality Check of RNA and Constructed Library

1. Quality check of isolated total RNA.
To assess the quality of the isolated total RNA, a simple run of an agarose gel could be used to check if the 28S and 16S rRNA bands are intact. However, a better assay would be to run an Agilent Bioanalyzer using a RNA Pico Chip and the result is shown in Fig. 2.

2. Quality check of fragmented cRNA.
To test the size distribution of RNA fragmentation, the Bioanalyzer traces of cRNA fragmentation as described in Subheading 3.4 is shown here (Fig. 3).

3. Quality check of the constructed DNA library.
The quality of the PAT-seq library produced can be examined by a number of ways. Here we give an example using a Agilent Bioanalyzer 2100 with a DNA Chip (Fig. 4).

4. Quantification of amplifiable cDNA library.
The effective concentration of the constructed cDNA library can be further examined by using the KAPA Illumina Library Quantification Kit. The result of this assay indicates how much of the DNA in the library contain both primers specific for Illumina sequencing cluster formation and annealing to the sequencing primers. Using the library shown in Fig. 3 as an example, the effective concentration is 17.82 nM. Libraries are adjusted to 10 nM according to recommendations from Illumina and then diluted to 2 nM working concentration for denaturing and sequencing.

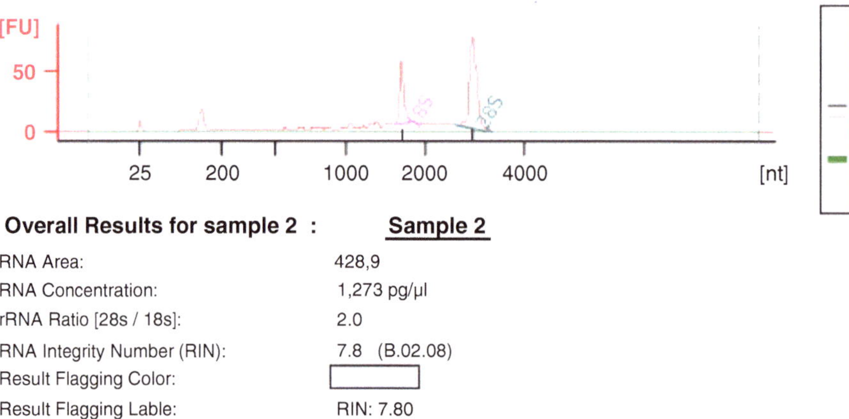

Fig. 2 Total RNA checked by Agilent RNA Pico Chip. *X* axis is the size of total RNA. *Y* axis is the fluorescent unit detected by Agilent Bioanalyzer 2100. The rRNA ratio is calculated by peak measurement of 28s to 18s. A gel image by the Agilent 2100 is given at the *right* with the *green band* indicating the front of gel and two rRNA bands. RNA Integrity Number (RIN) is an indication of the quality of the RNA, as good quality RNA is above 6.9 of RIN

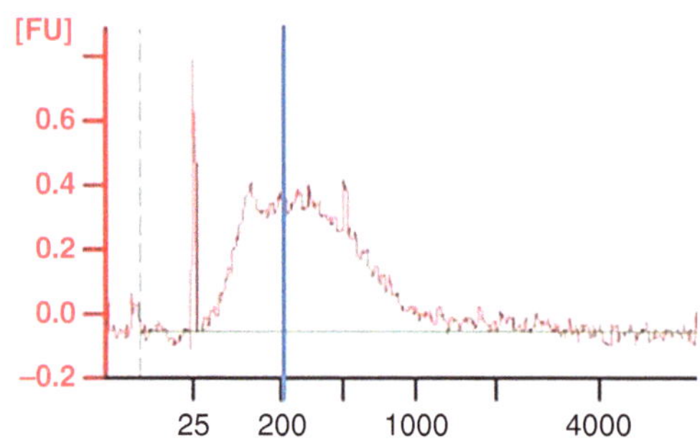

Fig. 3 cRNA fragmentation with first strand buffer. *X* axis presents the size distribution of cRNA incubated at 94 °C for 3 min. *Y* axis is the fluorescent unit detected by Agilent Bioanalyzer 2100. The *blue line* indicates 200 nt in length of fragmented cRNA

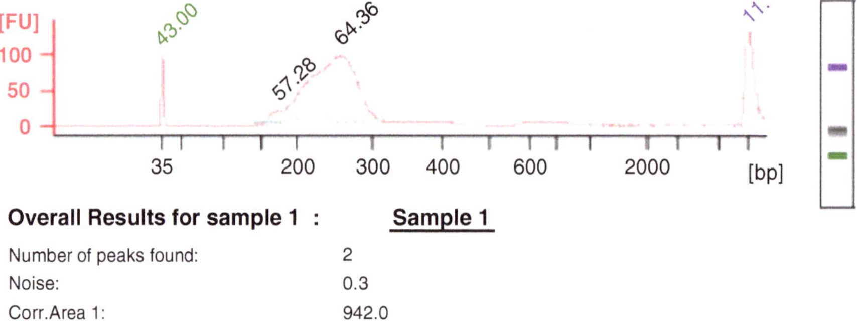

Fig. 4 The constructed cDNA library detected by Agilent high sensitivity DNA chip. *X* axis is the size of DNA library. *Y* axis is the fluorescent unit detected by Agilent Bioanalyzer 2100. A gel image by the Agilent 2100 is given at the *right* with the *green* and *purple bands* indicating the front and end of gel respectively, and the distribution of the DNA is indicated by the *black band*. This result suggests that the size and distribution of the library are in a desired range

4 Notes

1. Around 20,000 cells after cell sorting of *Arabidopsis thaliana* (ecotype Colombia) root tips were collected for total RNA isolation. The quality and quantity of isolated total RNA should be analyzed by Agilent Pico Chip. 10 ng of high quality total RNA (RIN $\geq$ 6.9, *see* Fig. 2) were used as starting material to construct the PAT-seq library.

2. Maximal T7 promoter sequence is used in T7-T18 primer in order to ensure adequate amplification by T7 RNA polymerase [11].

3. The 10× Reaction Buffer for T7 in vitro transcription has to be thawed at room temperature before the experiment set up. The spermidine in this buffer can come out of solution and co-precipitate the DNA if the reaction is prepared on ice.

4. When a new T7 in vitro transcription kit arrives, mix NTPs and make small aliquots of the mix and keep them in a –20 °C freezer. This is a good practice to avoid freeze-thaw cycles of NTPs.

5. For T7 in vitro transcription, increasing the temperature from 37 to 42 °C can increase the yield of amplified products. However, the incubation cannot exceed 4 h to avoid RNA degradation.

6. RNA fragmentation is carried out with divalent cations like magnesium in the first strand buffer under high heat [12]. The time for this incubation should be empirically determined due to variations of fragmentation efficiency (*see* Fig. 3). In order to reduce the loss of amplified products under high temperature, the shorter treatment is preferred.

7. The proper amount of RT-SE3n primer (*see* Table 1) with oligo d(T) to anneal at the 3′ end used in the second strand synthesis is to ensure enough end primer concentration for Klenow extension reaction [13].

8. Lower concentration of PE-PCR primers used is to avoid of self-annealing of the long adaptors, which will further block the DNA polymerase annealing on the cDNA templates. Single primer control in this PCR reaction should also be done to monitor potential faulty amplification.

9. <u>Barcoding</u> is introduced into the RT-SE primer, which is used to allow pooling different samples together in an Illumina sequencing lane. The two random nucleotides before the barcode are to increase heterogeneity in order to allow cluster recognition by Illumina sequencing protocols.

10. The italicized sequences in PE-PCR 1 and PE-PCR2 match the RT-SE and PE2-20-9N primers, respectively. This allows the PCR purification and the attachment of Illumina sequencing primers.

References

1. Xing D, Li QQ (2011) Alternative polyadenylation and gene expression regulation in plants. Wiley Interdiscip Rev RNA 2:445–458
2. Shen Y, Ji G, Haas BJ, Wu X, Zheng J, Reese GJ, Li QQ (2008) Genome level analysis of rice mRNA 3′-end processing signals and alternative polyadenylation. Nucleic Acids Res 36:3150–3161
3. Tian B, Hu J, Zhang H, Lutz CS (2005) A large-scale analysis of mRNA polyadenylation of human and mouse genes. Nucleic Acids Res 33:201–212
4. Mangone M et al (2010) The landscape of *C. elegans* 3′UTRs. Science 329:432–435
5. Wu X, Liu M, Downie B, Liang C, Ji G, Li QQ, Hunt AG (2011) Genome-wide landscape of polyadenylation in Arabidopsis provides evidence for extensive alternative polyadenylation. Proc Natl Acad Sci U S A 108:12533–12538
6. Di Giammartino DC, Nishida K, Manley JL (2011) Mechanisms and consequences of alternative polyadenylation. Mol Cell 43:853–866
7. Lutz CS (2008) Alternative polyadenylation: a twist on mRNA 3′ end formation. ACS Chem Biol 3:609–617
8. Zhang H, Lee J, Tian B (2005) Biased alternative polyadenylation in human tissues. Genome Biol 6:R100
9. Ara T, Lopez F, Ritchie W, Benech P, Gautheret D (2006) Conservation of alternative polyadenylation patterns in mammalian genes. BMC Genomics 7:189
10. Ohara R, Kikuno RF, Kitamura H, Ohara O (2005) CDNA library construction from a small amount of RNA: adaptor-ligation approach for two-round cRNA amplification using T7 and SP6 RNA polymerases. Biotechniques 38:451–458
11. Loudig O, Milova E, Brandwein-Gensler M, Massimi A, Belbin TJ, Childs G, Singer RH, Rohan T, Prystowsky MB (2007) Molecular restoration of archived transcriptional profiles by complementary-template reverse-transcription (CT-RT). Nucleic Acids Res 35:e94
12. Wang L, Si Y, Dedow LK, Shao Y, Liu P, Brutnell TP (2011) A low-cost library construction protocol and data analysis pipeline for Illumina-based strand-specific multiplex RNA-seq. PLoS One 6:e26426
13. Liu CL, Schreiber SL, Bernstein BE (2003) Development and validation of a T7 based linear amplification for genomic DNA. BMC Genomics 4:1–11

Chapter 17

A Rapid, Simple, and Inexpensive Method for the Preparation of Strand-Specific RNA-Seq Libraries

Arthur G. Hunt

Abstract

High-throughput sequencing of short cDNA tags, or RNA-Seq, has become a staple of genome-wide gene expression studies in plants. RNA-Seq libraries necessarily contain tags that correspond to the mRNA-poly(A) junction, or polyadenylation site, and thus may be mined for data that can help study alternative polyadenylation. This report presents a simple, rapid, and inexpensive method for preparing strand-specific RNA-Seq libraries from varying quantities of total RNA.

Key words Alternative polyadenylation, High-throughput sequencing, Gene expression, Strand-specific RNA-Seq

1   Introduction

Alternative polyadenylation, the process whereby different poly(A) sites in a pre-mRNA are utilized selectively or in regulated fashions, is an important mechanism for the regulation of gene expression. Besides the now well-established roles the process plays in the growth and differentiation of mammalian cells, it is also important for the control of gene expression in plants. Explicit links between alternative polyadenylation (or APA) and various processes in plants have been reported over the years (reviewed in [1]), and implicit connections with numerous other genes have appeared as well (e.g., [2]). For one model system, revolving around FLC expression and the control of flowering time in Arabidopsis, a wealth of mechanistic information is available [3]; in almost all other cases, the means by which APA in plants is initiated and controlled are not known.

For this reason, there is a need for the development of resources that can query poly(A) sites on a genome-wide scale in plants. Different approaches for the production, sequencing, and analysis of cDNA tags that specifically query the mRNA-poly(A) junction have been developed [4–6]; many of these are described in other

Arthur G. Hunt and Qingshun Quinn Li (eds.), *Polyadenylation in Plants: Methods and Protocols*, Methods in Molecular Biology, vol. 1255, DOI 10.1007/978-1-4939-2175-1_17, © Springer Science+Business Media New York 2015

chapters of this volume. Direct RNA sequencing has also been applied to the study of poly(A) site choice in plants [7–9]. These strategies have proven to be very useful, but have a limited focus and utility. A more widely used transcriptomic technology is RNA-Seq, the generation and sequencing of random cDNA tags that collectively query the complete transcriptome of an organism. RNA-Seq libraries necessarily include sequences that correspond to the mRNA-poly(A) junction, and thus may be a useful resource for the study of APA in various experimental systems. While this has not been implemented in plants, the utility of this approach is made clear from studies in which RNA-Seq libraries derived from various unicellular organisms have been mined for poly(A) site tags, and the resulting sequences used to assess APA [10–12].

The uses of RNA-Seq in plant science are manifest and varied (e.g., [13–15]), and the ongoing generation of RNA-Seq datasets should allow considerable data mining opportunities for the study of APA in plants. As the use of RNA-Seq grows, so too will the need for facile and inexpensive methods for the production of RNA-Seq libraries. The protocol described in this report is one such method. It is a variation of methods (e.g., [16]) that capitalize on unique features of reverse transcriptases derived from Moloney Murine Leukemia Viruses (MMuLVs; [17]); chief among these features is the propensity of the enzyme to append short, untemplated oligo-C tracts to the 3′ ends of cDNAs, and to switch templates using base pairing between these untemplated C's and 3′-terminal runs of G on the secondary template. The method detailed in this report has the following features: it generates strand-specific sequence tags without the need for modified nucleotide precursors, it is easily adaptable to different high-throughput sequencing platforms, it allows for extensive, user-defined barcoding and thus for multiplexing of samples at the sequencing step, and it is rapid and involves a minimal number of steps. Pertinent to the last point, modifications such as end-polishing, addition of untemplated tracts using reagents other than reverse transcriptase, and ligations are avoided. Importantly, the method in principle can be used to generate strand-specific RNA-Seq libraries from RNA isolated from bacteria as well as from eukaryotes. Because of these features, and because the cost per library is very low, the method is well suited for teaching as well as research purposes, in undergraduate [18] as well as graduate laboratory settings.

2 Materials and Equipment

500 μL thin-walled microcentrifuge tubes.

200 μL thin-walled PCR tubes.

Thermocyclers (set up to use either 200 μL tubes or 96-well microtiter plates).

Magnetic oligo-dT beads (New England Biolabs).

Magnetic stand for collecting oligo-dT beads.

RNAse-free water.

Binding Buffer for poly(A) enrichment: 20 mM Tris–HCl pH 7.5, 1.0 M LiCl, and 2 mM EDTA.

Wash Buffer for poly(A) enrichment: 10 mM Tris–HCl pH 7.5, 0.15 M LiCl, 1 mM EDTA.

10 mM Tris–HCl, pH 7.5.

SMARTSCRIBE Reverse Transcriptase (100 U/μL) and 5× First Strand Buffer (Clontech).

10 mM dNTP stock for reverse transcription reactions: 10 mM each of dATP, dGTP, dCTP, and dTTP.

100 mM DTT.

RNAse inhibitor (40 U/μL).

Primers for reverse transcription, strand-switching, and PCR— *see* Fig. 1.

AMPure beads (Beckman-Coulter).

80 % ethanol.

Phire Taq Polymerase (Thermo-Fisher).

2.5 mM dNTP stock for PCR reactions: 2.5 mM each of dATP, dGTP, dCTP, and dTTP.

Apparatus for agarose gel electrophoretic separation and visualization of DNA.

Agarose.

1× TBE: 100 mM Tris–boric acid, pH 8.0, 1 mM Na EDTA.

3 Protocol

3.1 Overview

The protocol for preparing strand-specific RNA-Seq libraries is illustrated in Fig. 2, and may be divided into three steps: RNA preparation, cDNA synthesis, and PCR amplification. The protocol as described is intended for the study of eukaryotic mRNA, and thus includes a poly(A) enrichment step. However, alternative RNA preparations (such as removal of ribosomal RNA from eukaryotic or prokaryotic samples) may also be used.

3.2 Poly(A) Enrichment (*See* Note 1)

1. Aliquot 15 μL of oligo(dT) beads (NEB) to a thin-walled 500 μL microcentrifuge tube (*see* **Note 2**). Add 100 μL of Binding Buffer, mix, and collect the beads using the magnetic stand. Remove the supernatant and repeat the wash. After removing the supernatant, resuspend the beads in 50 μL of Binding Buffer.

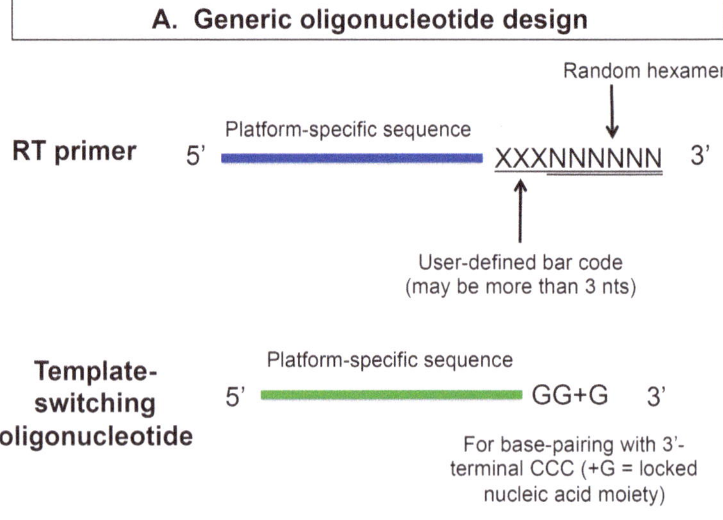

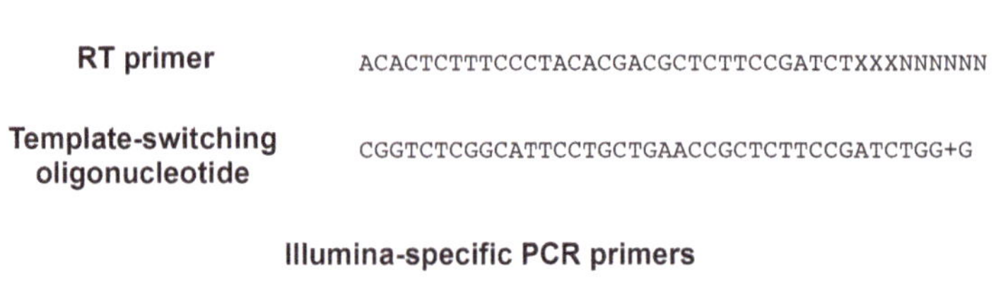

Fig. 1 Illustration of the primers used to prepare RNA-seq libraries for high-throughput sequencing. (**a**) Generic oligonucleotide design, showing portions of each primer and corresponding function. (**b**) Sequences of the oligonucleotides used to generate Illumina-specific libraries. The RT and template-switching primers are shown at the *top*, and the two primers used for PCR amplification of tags beneath. "1" denotes the primer that directs the first set of Illumina sequencing reads. "2" denotes the primer that directs the second set of reads in instances where paired-end sequencing is desired. As illustrated, single-end reads will possess the bar code and thus are sufficient for a routine RNA-Seq study

2. Mix total RNA (*see* **Note 3**) and RNAse-free water to a total volume of 50 μL. Incubate the diluted RNA at 65 °C for 5 min. Place tube on ice.

3. Add the entire RNA sample to the beads from **step 1**. Incubate at room temperature for 5 min.

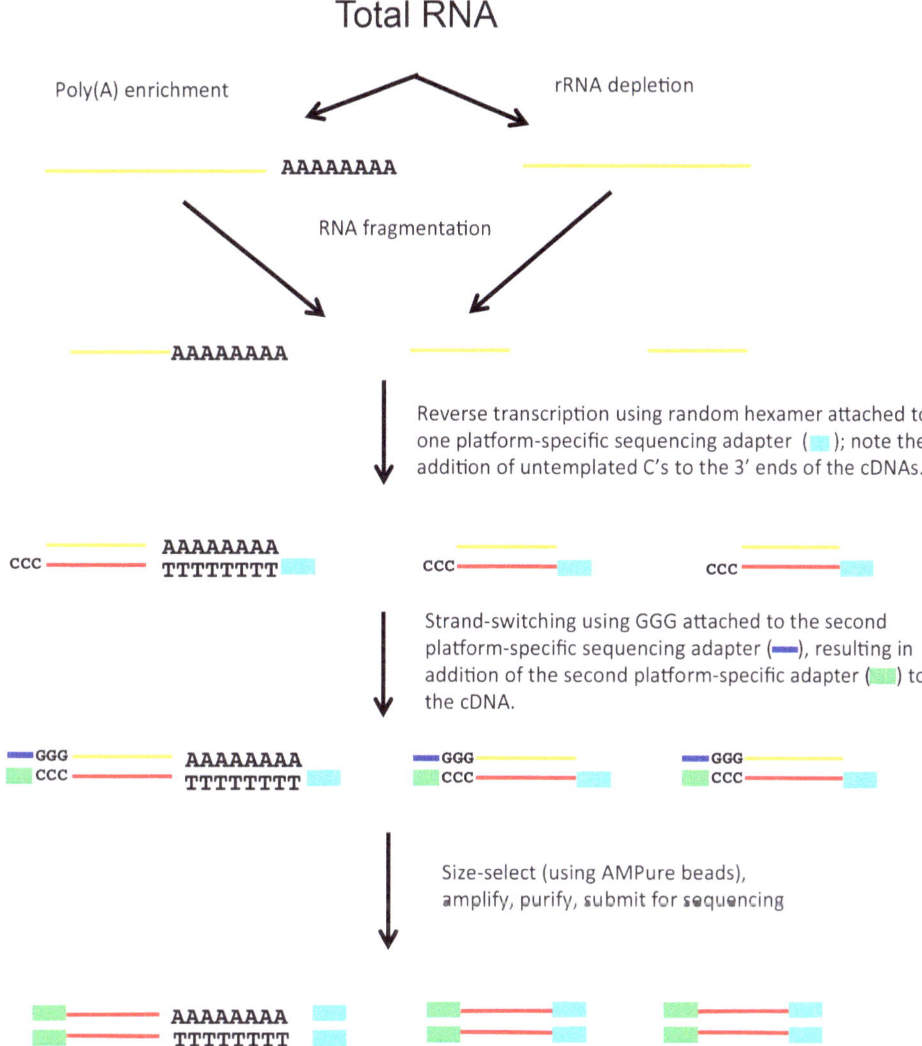

Fig. 2 Illustration of the steps involved in preparation of RNA-Seq libraries. Total RNA from either eukaryotic or prokaryotic sources may be used; thus, as illustrated for the first step, poly(A) enrichment or rRNA depletion may be used to process the RNA before fragmentation and cDNA synthesis

4. Collect the beads using the magnetic stand and remove the supernatant. Add 100 μL of Wash Buffer. Mix, collect the beads with the magnetic stand, and remove and discard the supernatant. Repeat and save the beads for the next step.

5. To the pelleted beads from **step 4**, add 15 μL of 10 mM Tris–HCl, pH 7.5. Suspend the beads and heat to 80 °C. After 3 min, remove the beads with the magnetic stand and quickly collect the supernatant (*see* **Notes 4–6**).

3.3 cDNA Synthesis and Cleanup

1. Mix 14.5 μL RNA, 1 μL (=100 pmol) RT primer (*see* **Note 7**) + 5 μL 5× first strand buffer.

2. Heat to 95 °C for 2 min, chill (*see* **Note 8**).

3. *Immediately* add (while cold!):

   2.5 μL 10× dNTPs for reverse transcription reactions.

   1 μL 100 mM DTT.

   0.5 μL RNase Inhibitor.

   1 μL SMARTSCRIBE.

   (*See* **Note 9**).

4. Mix and incubate for 120 min at 42 °C (*see* **Note 10**).

5. Add 1 μL of the template-switching oligonucleotide and an additional 1 μL of SMARTSCRIBE, mix, and incubate for an additional 120 min at 42 °C (*see* **Note 11**).

6. Heat to 70 °C for 5 min.

7. Add 25 μL AMPure beads that have been completely mixed and brought to room temperature. Incubate for 8 min at room temperature.

8. Separate beads using the magnetic stand (this may take a minute or two, because of the viscosity of the bead solution), and remove and discard the supernatant.

9. Add 200 μL *fresh* 80 % ethanol (*see* **Note 12**), mix, and collect the beads. Remove and discard the supernatant. Repeat the wash.

10. Air-dry the washed beads for 5 min at room temperature.

11. Add 25 μL 10 mM Tris–HCl, pH 7.5. Mix, and then collect the beads with the magnetic stand. Remove the supernatant to a new 500 μL thin-walled microcentrifuge tube.

12. Repeat **steps 7–11**. Save the final 25 μL elution for subsequent PCR reactions (*see* **Note 6**).

3.4 PCR Amplification of Tags

1. Prepare a master mix for PCR reactions: for each reaction, mix 5 μL of 5× Phire reaction buffer (*see* **Note 13**), 2.5 μL 10× dNTPs for PCR, 10–20 pmol of each primer (PE-PCR1 and PE-PCR2; Fig. 1), 0.5 μL Phire Taq polymerase, and water to bring the volume to 24 μL. (For example, for ten reactions, mix 50 μL of 5× Phire reaction buffer, 25 μL of 2.5 mM dNTPs, 100 pmol of each primer, 5 μL of Phire Taq DNA polymerase, and water to bring the final volume to 240 μL.)

2. Aliquot 1 μL (*see* **Note 14**) of the cDNA from Subheading 3.3, **step 12** to a 200 μL thin-walled PCR tube. Add 24 μL of the master mix from **step 1**, mix, and execute the following PCR program: 3 min at 98 °C, between 9 and 20 cycles each of

which consists of 15 s at 98 °C, 15 s at 60 °C, and 1 min at 72 °C, and a final extension at 72 °C for 10 min (*see* **Note 15**).

3. Assess the amplicons on an agarose gel. Select the optimal conditions (*see* **Note 16**) and repeat the amplification as in **step 2**.

4. Add 25 µL AMPure beads that have been completely mixed and brought to room temperature. Incubate for 8 min at room temperature.

5. Separate beads using the magnetic stand (this may take a minute or two, because of the viscosity of the bead solution), and remove and discard the supernatant.

6. Add 200 µL *fresh* 80 % ethanol (*see* **Note 12**), mix, and collect the beads. Remove and discard the supernatant. Repeat the wash.

7. Air-dry the washed beads for 5 min at room temperature.

8. Add 25 µL 10 mM Tris–HCl, pH 7.5. Mix, and then collect the beads with the magnetic stand. Remove the supernatant to a new 500 µL thin-walled microcentrifuge tube. This is the sample to be submitted for high-throughput sequencing (*see* **Note 16**).

4   Notes

1. This protocol is presented for use with eukaryotic, polyadenylated RNA, as the focus of the volume is the study of polyadenylation in plants. However, prokaryotic RNA can be used for library preparation. In such cases, one or two rounds of ribosomal RNA removal are recommended before proceeding. Also, Subheading 3.2 may be omitted, and the user may begin with Subheading 3.3.

2. The quantity of oligo-dT beads can be varied, depending on the supplier and poly(A) RNA-binding capacity of the beads.

3. Any suitably purified RNA can be used for the library preparation. Intact RNA may be preferable, but RNA of lesser quality can probably be used as well. This follows from the realization that the RNA is deliberately fragmented in the course of the library preparation.

4. The process of moving the microcentrifuge tube from the 80 °C heat block to the magnetic stand, and subsequent recovery of the eluted RNA into a new tube, should take no more than 10 s. This is to minimize rebinding of the polyadenylated RNA to the beads once the tube has cooled.

5. Depending on the source and nature of the total RNA sample, it may be necessary to repeat the poly(A) enrichment with the sample recovered in Subheading 3.2, **step 5**.

6. These are steps where the procedure may be halted (as at the end of a workday, or the end of a teaching laboratory period). RNA should be stored at −80 °C, and cDNA may be stored at −20 °C.

7. The primers used for reverse transcription reactions may be adapted for any high-throughput DNA sequencing platform; those shown in Fig. 1 are designed to generate DNA tags for sequencing on Illumina (GAIIx, HiSeq, MiSeq) platforms. In a generic sense, these primers have three parts—a sequence situated at the 5′ end that is specific for the sequencing platform, a bar code of user-defined length and sequence (the primer family shown in Fig. 1 has a 3 nt bar code), and a 3′-terminal random hexamers. Typically, the platform-specific part of the sequence will be such that the first base of the sequence will be the first base of the bar code. It should be noted that the platform-specific portion of this primer will correspond to the first sequencing primer in situations where paired-end sequencing is desired. By using one primer (and not both), this protocol inevitably yields strand-specific RNA-Seq tags.

8. Subheading 3.3, **step 2** serves two purposes: it melts secondary structure in the RNA population, and it promotes a limited fragmentation of the RNA. The elimination of secondary structure promotes subsequent primer annealing and passage of reverse transcriptase. The fragmentation serves to promote the production of DNA tags of a size suited for sequencing. The time of fragmentation may be varied so as to promote shorter or longer tags, as desired by the user.

9. For multiple samples, prepare a master mix for $n+1$ samples, and keep this mix on ice. This minimizes pipetting steps and the attendant possibility of errors.

10. The time and temperature recommended in Subheading 3.3, **step 4** are selected to promote optimal yields. Long times (2 h or more) serves to permit the completion of long cDNAs, and to also allow reasonable cDNA yields under conditions (such as when small quantities of starting RNA are used) in which the concentration of RNA template is substantially lower than the affinity of the enzyme for template. The choice of 42 °C as the reaction temperature reflects the recommendation of the supplier of the reverse transcriptase for use with random primed cDNA synthesis.

11. The design of the strand-switching primer is informed by the reported properties of the reverse transcriptase and studies that

reveal enhanced performance of strand-switching primers that have at least one 3′-terminal locked nucleic acid [19, 20]. MMuLV-derived reverse transcriptases are known to add untemplated tracts of dC to the 3′-ends of cDNAs formed by complete copying of the template RNA [17]. Accordingly, the strand-switching primer will possess, at its very 3′-end, a GGG sequence intended to base-pair with the 3′-terminal CCC of the cDNA. The 3′-terminal locked nucleic acid facilitates this base-pairing at the temperature used for the reverse transcriptase reactions. While all possible permutations have not been evaluated, the order of operations described here (a 2-h reverse transcription reaction performed in the absence of a strand-switching primer, followed by addition of the second primer and of an additional aliquot of reverse transcriptase) is based on a systematic study of the strand-switching step of the reaction [21].

12. It is important that the 80 % ethanol be freshly prepared, since it is hygroscopic and will acquire a greater water content upon storage.

13. The use of a hot-start thermostable DNA polymerase is highly recommended, so as to minimize the generation of primer dimers and other artifacts at the beginning of the PCR process. The conditions described here are optimized for use with the Phire enzyme from Thermo-Fisher. However, other hot-start enzymes may also be used. In such instances, it may be necessary to optimize annealing temperatures and extension times to promote optimal cDNA yields.

14. The quantities of cDNA template may be varied such that as many as 10 μL may be used. In such instances, the volume of water used to prepare the master mix described in Subheading 3.4, **step 1**, will have to be adjusted accordingly.

15. The number of cycles for tag amplification may be varied between 9 and 20, depending on the quantities of starting RNA. The desired result is a smear of DNA ranging in size from about 200 bp to 500 bp (*see* Fig. 3). The use of more than 25 cycles is not recommended, as this usually leads to results in which single low-abundance cDNAs are artificially overrepresented in the final sequencing outcomes. Samples comparable to lanes 1–4 have been successfully purified using AMPure and sequenced. Note that, in Fig. 3, there is an approximately three-cycle difference in the "yields" of tags derived from libraries prepared with 1,000 and 100 ng of total Arabidopsis RNA, respectively. This reflects the direct correlation between input RNA and tag yield.

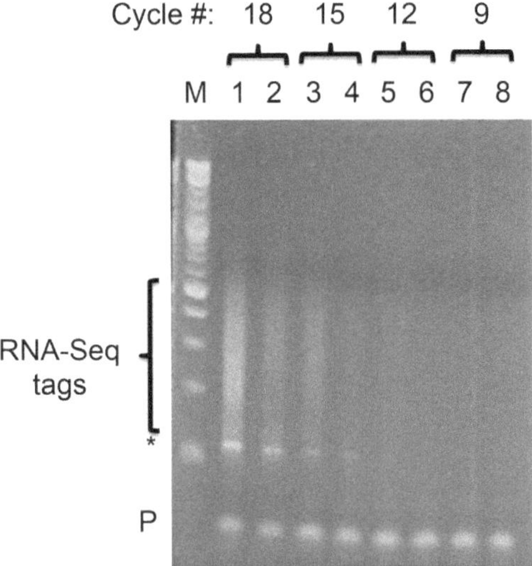

Fig. 3 Typical outcomes for RNA-Seq library preparations. In this experiment, RNA-Seq libraries were made beginning with 1,000 (lanes 1, 3, 5, and 7) or 100 (lanes 2, 4, 6, and 8) ng of total RNA isolated from Arabidopsis cell cultures. The libraries were prepared using a single round of poly(A) enrichment (Subheading 3.2). 1 µL of the cDNA (from Subheading 3.3, **step 12**) was used for amplifications as described (Subheading 3.4, **step 2**) for the numbers of cycles indicated above the gel image. Amplicons were separated on a 1.5 % agarose gel cast and run in 1× TBE

16. A representative set of results is shown in Fig. 4. Figure 4a depicts the bioanalyzer traces from tags amplified as shown in Fig. 3, lanes 3 (=sample 100 in Fig. 4a) and 2 (=sample 101 in Fig. 4a); these runs were performed at the Advanced Genetic Technologies Center (AGTC) at the University of Kentucky. Of note is the size distribution of the tags, reflecting the effectiveness of the fragmentation and random distribution of the RT priming. Figure 4b shows the results of mapping of the tags from samples 100 and 101 onto the Arabidopsis genome. For this, the tags shown in Fig. 4a were sequenced on a MiSeq instrument at AGTC. After de-multiplexing, approximately three million reads for each sample were returned. This initial sequence processing, as well as the subsequent mapping and analysis, was performed using CLC Genomics Workbench (Version 7.0.3). Mapping to the Arabidopsis genome was performed using the Large-Gap mapping tool and the default settings (length fraction = 0.5, sequence identity = 0.8). Mapping results were displayed as shown using the Tracks tools in CLC Genomics Workbench.

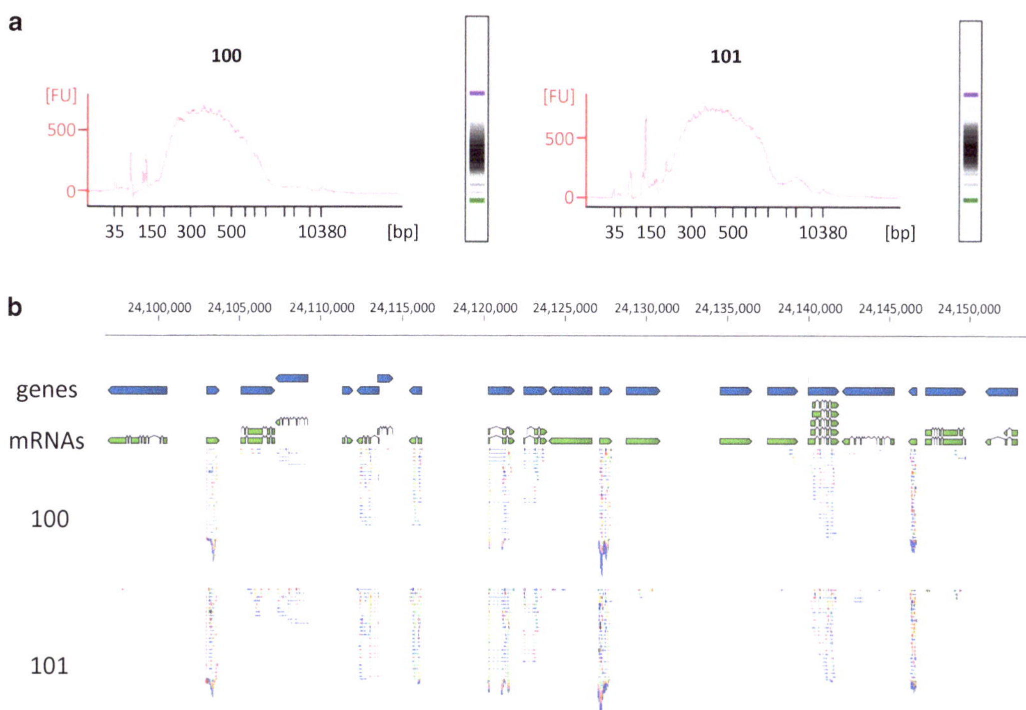

Fig. 4 (a) Bioanalyzer traces for the samples illustrated in lanes 2 and 3 (numbered as 101 and 100, respectively) in Fig. 3. The *x*-axes for each plot are a log10 representation of DNA size standards, the sizes of which are shown below the axes. The *y*-axes are arbitrary units that reflect DNA abundance. At the *right* of each graph is shown a photograph of the bioanalyzer run; this was generated by the instrument. (b) Representative results of the mapping of reads from samples 100 and 101 to Arabidopsis chromosome 1. Sequencing and analyses were performed as described in **Note 16**. The plot was generated using the Track Tools package of CLC Genomics Workbench. Four tracks are shown beneath the line designating chromosome coordinates; these are as indicated with the labels on the *left*. The "genes" track illustrates the genomic locations of annotated genes in this chromosomal region. The "mRNA" track shows the locations of mRNAs (and, consequently, of exons and introns); genes with more than one annotated mRNA can be seen in this track. The other two tracks show the mapping results for the respective samples; *thick horizontal tics* represent individual reads, and *thinned* or *dashed lines* between the tics denote gaps between paired-ends or reads that span introns or other RNA processing events. *Blue tics* represent reads in which both pairs map, while *green* and *red tics* denote mappings of broken pairs. *Yellow tics* denote reads that map to more than one genomic location (color figure online)

Acknowledgements

This work was supported by the National Science Foundation (awards IOS-0817818 and MCB-1243849). The author thanks the staff of AGTC for many helpful discussions and for much patience, and Carol Von Lanken for excellent technical and administrative assistance.

References

1. Hunt AG (2008) Messenger RNA 3′ end formation in plants. Curr Top Microbiol Immunol 326:151–177

2. Tan X, Meyers BC, Kozik A, West MA, Morgante M, St Clair DA, Bent AF, Michelmore RW (2007) Global expression analysis of nucleotide binding site-leucine rich repeat-encoding and related genes in Arabidopsis. BMC Plant Biol 7:56. doi:10.1186/1471-2229-7-56

3. Rataj K, Simpson GG (2014) Message ends: RNA 3′ processing and flowering time control. J Exp Bot 65(2):353–363. doi:10.1093/jxb/ert439

4. Ma L, Pati PK, Liu M, Li QQ, Hunt AG (2013) High throughput characterizations of poly(A) site choice in plants. Methods. doi:10.1016/j.ymeth.2013.06.037

5. Thomas PE, Wu X, Liu M, Gaffney B, Ji G, Li QQ, Hunt AG (2012) Genome-wide control of polyadenylation site choice by CPSF30 in Arabidopsis. Plant Cell 24(11):4376–4388. doi:10.1105/tpc.112.096107

6. Wu X, Liu M, Downie B, Liang C, Ji G, Li QQ, Hunt AG (2011) Genome-wide landscape of polyadenylation in Arabidopsis provides evidence for extensive alternative polyadenylation. Proc Natl Acad Sci U S A 108(30): 12533–12538. doi:10.1073/pnas.1019732108

7. Duc C, Sherstnev A, Cole C, Barton GJ, Simpson GG (2013) Transcription termination and chimeric RNA formation controlled by Arabidopsis thaliana FPA. PLoS Genet 9(10):e1003867. doi:10.1371/journal.pgen.1003867

8. Lyons R, Iwase A, Gansewig T, Sherstnev A, Duc C, Barton GJ, Hanada K, Higuchi-Takeuchi M, Matsui M, Sugimoto K, Kazan K, Simpson GG, Shirasu K (2013) The RNA-binding protein FPA regulates flg22-triggered defense responses and transcription factor activity by alternative polyadenylation. Sci Rep 3:2866. doi:10.1038/srep02866

9. Sherstnev A, Duc C, Cole C, Zacharaki V, Hornyik C, Ozsolak F, Milos PM, Barton GJ, Simpson GG (2012) Direct sequencing of Arabidopsis thaliana RNA reveals patterns of cleavage and polyadenylation. Nat Struct Mol Biol 19(8):845–852. doi:10.1038/nsmb.2345

10. Janbon G, Ormerod KL, Paulet D, Byrnes EJ 3rd, Yadav V, Chatterjee G, Mullapudi N, Hon CC, Billmyre RB, Brunel F, Bahn YS, Chen W, Chen Y, Chow EW, Coppee JY, Floyd-Averette A, Gaillardin C, Gerik KJ, Goldberg J, Gonzalez-Hilarion S, Gujja S, Hamlin JL, Hsueh YP, Ianiri G, Jones S, Kodira CD, Kozubowski L, Lam W, Marra M, Mesner LD, Mieczkowski PA, Moyrand F, Nielsen K, Proux C, Rossignol T, Schein JE, Sun S, Wollschlaeger C, Wood IA, Zeng Q, Neuveglise C, Newlon CS, Perfect JR, Lodge JK, Idnurm A, Stajich JE, Kronstad JW, Sanyal K, Heitman J, Fraser JA, Cuomo CA, Dietrich FS (2014) Analysis of the genome and transcriptome of Cryptococcus neoformans var. grubii reveals complex RNA expression and microevolution leading to virulence attenuation. PLoS Genet 10(4):e1004261. doi:10.1371/journal.pgen.1004261

11. Schlackow M, Marguerat S, Proudfoot NJ, Bahler J, Erban R, Gullerova M (2013) Genome-wide analysis of poly(A) site selection in Schizosaccharomyces pombe. RNA 19(12):1617–1631. doi:10.1261/rna.040675.113

12. Zhao Z, Wu X, Kumar PK, Dong M, Ji G, Li QQ, Liang C (2014) Bioinformatics analysis of alternative polyadenylation in green alga Chlamydomonas reinhardtii using transcriptome sequences from three different sequencing platforms. G3 (Bethesda) 4(5):871–883. doi:10.1534/g3.114.010249

13. Jimenez-Gomez JM (2011) Next generation quantitative genetics in plants. Front Plant Sci 2:77. doi:10.3389/fpls.2011.00077

14. O'Rourke JA, Bolon YT, Bucciarelli B, Vance CP (2014) Legume genomics: understanding biology through DNA and RNA sequencing. Ann Bot 113(7):1107–1120. doi:10.1093/aob/mcu072

15. Strickler SR, Bombarely A, Mueller LA (2012) Designing a transcriptome next-generation sequencing project for a nonmodel plant species. Am J Bot 99(2):257–266. doi:10.3732/ajb.1100292

16. Ramskold D, Luo S, Wang YC, Li R, Deng Q, Faridani OR, Daniels GA, Khrebtukova I, Loring JF, Laurent LC, Schroth GP, Sandberg R (2012) Full-length mRNA-Seq from single-cell levels of RNA and individual circulating tumor cells. Nat Biotechnol 30(8):777–782. doi:10.1038/nbt.2282

17. Zhu YY, Machleder EM, Chenchik A, Li R, Siebert PD (2001) Reverse transcriptase template switching: a SMART approach for full-length cDNA library construction. Biotechniques 30(4):892–897

18. Buonaccorsi V, Peterson M, Lamendella G, Newman J, Trun N, Tobin T, Aguilar A, Hunt A, Praul C, Grove D, Roney J, Roberts W

(2014) Vision and change through the genome consortium for active teaching using next-generation sequencing (GCAT-SEEK). CBE Life Sci Educ 13(1):1–2. doi:10.1187/cbe.13-10-0195

19. Picelli S, Bjorklund AK, Faridani OR, Sagasser S, Winberg G, Sandberg R (2013) Smart-seq2 for sensitive full-length transcriptome profiling in single cells. Nat Methods 10(11):1096–1098. doi:10.1038/nmeth.2639

20. Picelli S, Faridani OR, Bjorklund AK, Winberg G, Sagasser S, Sandberg R (2014) Full-length RNA-seq from single cells using Smart-seq2. Nat Protoc 9(1):171–181. doi:10.1038/nprot.2014.006

21. Pinto FL, Lindblad P (2010) A guide for in-house design of template-switch-based 5′ rapid amplification of cDNA ends systems. Anal Biochem 397(2):227–232. doi:10.1016/j.ab.2009.10.022

Chapter 18

Genome-Wide Analysis of Distribution of RNA Polymerase II Isoforms Using ChIP-Seq

Laura de Lorenzo

Abstract

Chromatin immunoprecipitation followed by sequencing (ChIP-seq) is a powerful technique for genome-wide profiling of DNA-binding proteins in vivo. ChIP has been used to study diverse nuclear processes such as transcription regulation, at specific loci as well as across the entire genome. In this report, a protocol is described for the application of ChIP to the genome-wide analysis of the distribution of different RNA polymerase II forms. The method makes use of the possibility to crosslink proteins to the DNA, to which they bind in vivo. Specific RNA-Pol II–DNA complexes can then be purified by immunoprecipitation using a specific antibody against the DNA-binding protein of interest, and the associated DNA fragments recovered and analyzed.

Key words Chromatin immunoprecipitation (ChIP), RNA polymerase II, Gene expression, Next-generation sequencing, ChIP-seq

1 Introduction

Along with the canonical collection of subunits that make up the core polyadenylation complex, other proteins play key roles in the process of mRNA 3′ end formation in eukaryotes. Among these are RNA polymerase II, through the functions of the distinctive C-terminal domain of the protein [1]. This C-terminal domain (CTD) is composed of consecutive repeats of the sequence Tyr-Ser-Pro-Thr-Ser-Pro-Ser $(Y_1S_2P_3T_4S_5P_6S_7)_n$ and the main stages of the transcription process are associated with distinct phosphorylation and dephosphorylation states of CTD [2–4]. Specifically, the non-phosphorylated form of Pol II is involved in the formation of the preinitiation complex (PIC); phosphorylation of Ser7-CTD (Ser7P-CTD) is placed early in transcription [5]; Ser5P-CTD participates in transcription initiation and early elongation, and the Ser2P-CTD modification recruits factors for mRNA polyadenylation and termination [6, 7]. Understanding the nature of CTD modifications, including how they function and how they are regulated, is central

Arthur G. Hunt and Qingshun Quinn Li (eds.), *Polyadenylation in Plants: Methods and Protocols*, Methods in Molecular Biology, vol. 1255, DOI 10.1007/978-1-4939-2175-1_18, © Springer Science+Business Media New York 2015

for knowing the mechanisms that control gene expression. This is especially true for the study of mRNA polyadenylation and transcription termination.

The impact that chromatin immunoprecipitation (ChIP) has had on the field of nuclear signaling and transcriptional control is unquestionable. It has tremendously advanced our understanding of the action of transcription factors, cofactors and histone modifications on the regulation of gene expression [8, 9]. Different protocols have been proposed to study the protein–DNA interaction using ChIP coupled to different high-throughput technologies [10–12]. Here, a ChIP protocol is described that may be coupled to deep sequencing (ChIP-seq) and is suitable for the study of the genome-wide distribution of the different phosphorylated forms of Pol II using highly specific antibodies, each directed against one of the different phosphorylated forms of CTD. This protocol is efficient and reproducible and may be used to determine how much RNA polymerase II is bound to the DNA at the different stages of the transcription process, at individual genes and genome-wide.

2   Materials

Prepare all solutions using ultrapure water (obtained by purifying deionized water to attain a sensitivity of 18 MΩ cm at 25 °C) and analytical grade reagents. Prepare and store all reagents at room temperature (unless otherwise indicated). All waste disposal regulations are strictly followed when disposing waste materials.

2.1   Equipment

Mortar and pestle.

- DNA Lo-bind tubes.
- Standard vacuum pump.
- Sonicator (e.g., Bioruptor® UCD-200, Bioruptor NGS, Diagenode).
- Refrigerated centrifuges for 1.5 and 50 mL tubes.
- Standard and real-time PCR machines.
- Rotating wheel.
- Magnetic separation rack for use with microcentrifuge tubes.
- Bioanalyzer (Agilent Technologies; Model 2100, high sensitive DNA chip).

2.2   Reagents

Plant material (fresh tissue, *see* **Note 1**).

- Purified antibodies (1–2 µg per ChiP experiment) (*see* **Note 2**).
- Liquid nitrogen.
- Isopropyl alcohol.

- Miracloth (Calbiochem) or 44-μm mesh.

- MinElute PCR purification kit to purify double-stranded DNA fragments from enzymatic reactions (Qiagen) and MinElute gel extraction kit (Qiagen).

- Proteinase K (20 mg/mL).

- RNase A/T1 (10 mg/mL).

- Complete protease inhibitor cocktail tablets, EDTA-free (Roche): complete ULTRA tablet, EDTA-free or/and complete MINI tablet, EDTA-free (*see* **Note 3**).

- Phosphatase Inhibitor Cocktail Tablets (PhosSTOP, Roche).

- Protein-A magnetic beads (New England Biolabs).

- 1 M Tris–HCl pH 9.

- TE buffer—10 mM Tris–HCl, pH 8.0, 1 mM Na EDTA.

- 3 M sodium acetate pH 5.2.

- 37 % formaldehyde.

- 1.25 M glycine.

- iQ SYBR Green Supermix (Bio-Rad).

- Phusion DNA polymerase (Thermo Scientific).

- 100 bp ladder.

- AMPure® XP Beads (Beckman Coulter, Inc.).

- NEBNext® Ultra™ DNA Library Prep Kit for Illumina or NEBNext® ChIP-Seq Library Prep Reagent Set for Illumina.

- NEBNext Singleplex or Multiplex Oligos for Illumina

- *MC buffer* (*fresh*): 10 mM sodium phosphate pH 7.0, 50 mM NaCl, and 0.1 M sucrose. To prepare 1 M sodium phosphate pH 7.0, mix 57.7 mL 1 M disodium hydrogen phosphate (Na_2HPO_4) and 42.3 mL 1 M sodium dihydrogen phosphate (NaH_2PO_4).

- *M1 buffer* (*fresh*): 10 mM sodium phosphate pH 7.0, 0.1 M NaCl, 1 M 2-methyl 2,4-pentanediol, 10 mM β-mercaptoethanol, and complete protease inhibitor cocktail (*see* **Note 3**).

- *M2 buffer* (*fresh*): M1 buffer + 10 mM $MgCl_2$ and 0.5 % Triton X-100.

- *M3 buffer* (*fresh*): M1 buffer, but without 2-methyl 2,4-pentanediol.

- *Sonication buffer* (*fresh or store frozen in aliquots at −20°C*): 10 mM sodium phosphate pH 7.0, 0.1 M NaCl, 0.5 % sarkosyl, 10 mM EDTA, 1 mM 4-(2-Aminoethyl)benzenesulfonyl fluoride hydrochloride (Pefabloc SC, from Sigma Aldrich), and complete protease inhibitor cocktail (*see* **Note 3**).

- *IP buffer (fresh or store frozen in aliquots at −20°C):* 50 mM HEPES pH 7.5, 150 mM NaCl, 5 mM $MgCl_2$, 10 µM $ZnSO_4$, 1 % Triton X-100, and 0.05 % SDS.

- *Elution buffer (fresh or store frozen in aliquots at −20°C):* 0.1 M glycine, 0.5 M NaCl, 0.05 % Tween-20, adjusted to pH 2.8 with HCl.

- 80 % ethanol, prepared freshly (no more than 60 min before use).

3   Methods

The ChIP protocol is summarized in Fig. 1 and starts with tissue fixation (Subheading 3.1), followed by nuclei isolation (Subheading 3.2), immunoprecipitation (Subheading 3.3), ChIP wash and chromatin elution (Subheading 3.4), DNA purification and analysis (Subheading 3.5), and high-throughput sequencing (Subheading 3.6). As indicated in the following, certain parameters (such as formaldehyde concentration and incubation times as well as chromatin shearing) will have to be empirically determined for different experiments and samples. Regardless, during execution of the protocol, the samples should be kept on ice (unless indicated otherwise).

3.1 Tissue Collection and Fixation

1. Harvest 0.8–2 g of *Arabidopsis* seedlings in 25 mL cold MC Buffer (use 50 mL falcon tube), and taking care to keep the collected tissue cool (*see* **Note 4**).

2. Add fresh formaldehyde (37 %) directly to the MC buffer containing plant material to reach a final concentration of 1 % and incubate on ice under vacuum infiltration for 1 h (*see* **Note 5**).

3. Quench the crosslinking by adding 2.5 mL of 1.25 M glycine (final concentration 0.125 M), mix well by inverting the tube several times, and infiltrate under vacuum for an additional 5 min.

4. Dispose of the solution containing formaldehyde and glycine, rinse the seedlings three times with Milli-Q (deionized) water, remove as much water as possible using tissue paper, transfer the plant material to a new 50 mL tube, and flash freeze in liquid nitrogen (*see* **Note 6**).

3.2 Nuclei Isolation and Shearing of Chromatin

1. Grind the frozen tissue to a fine powder using a mortar and pestle, keeping the tissue frozen throughout.

2. Transfer the powder to a 50 mL falcon tube containing 20 mL cold M1 buffer (vortex to mix) or to an empty falcon tube to keep the powder at −80 °C (if samples need to be stored—*see* **Note 6**).

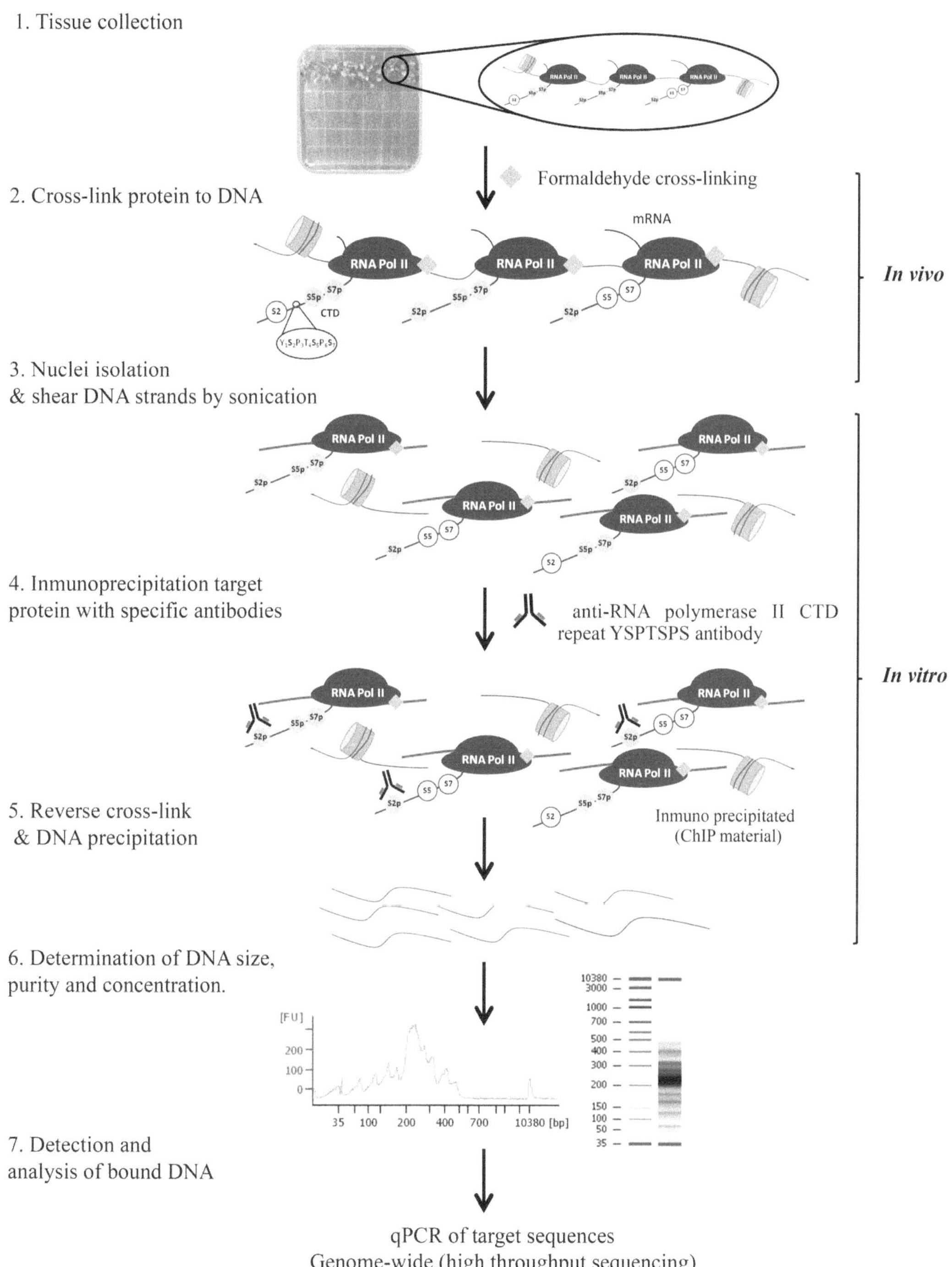

Fig. 1 A schematic representation of the chromatin immunoprecipitation (ChIP) assay. After a suitable growth and treatment (1), DNA and proteins are covalently cross-linked together using formaldehyde (2) and the chromatin is fragmented by sonication (3). Specific antibodies are used to immunoprecipitate the targets fragments of DNA (4). Once the cross-links are reversed (5), the DNA can be subjected to a number of downstream analysis techniques (6-7), such as quantitative PCR and genome-wide analyses using deep sequencing (ChIP-seq). An example of a ChIP-seq library prepared using anti-RNA polymerase II CTD repeat YSPTSPS antibody is shown at the *bottom*; the trace of a Bioanalyzer output is shown on the *left*, and the image of the run on the *right*. These results depict a typical library and represent a reasonable goal

3. Filter the slurry through two layers of Miracloth or 44-μm mesh and collect in a fresh 50 mL falcon tube on ice (filter three times using the same mesh). Wash the mesh with an additional 5 mL of M1 buffer to collect all nuclei in the filtrate (*see* **Note 7**).

4. Centrifuge the filtrate at $1,000 \times g$ for 20 min at 4 °C. Remove the supernatant.

5. Resuspend the nuclear pellet with 4 mL of cold M2 buffer, and centrifuge at $1,000 \times g$ for 5 min at 4 °C. Remove the supernatant. Repeat for a total of five washes.

6. Resuspend the pellet with 1 mL of cold M3 buffer, transfer the solution to a new 2 mL microcentrifuge tube, and spin down as in **step 5**. Remove the supernatant.

7. Resuspend the crude nuclear pellet in 1 mL Sonication buffer and transfer to a tube appropriate for sonication.

8. Shear the chromatin with a sonicator (*see* **Note 8**). Using the appropriate settings for sonication, the chromatin fragments should have sizes between 200 and 800 bp (*see* **Note 9**).

9. Centrifuge the suspension in a microcentrifuge at top speed for 10 min at 4 °C to pellet the insoluble material (*see* **Note 10**).

10. Transfer the solubilized chromatin to a new 2 mL microcentrifuge tube, spin 10 min at top speed at 4 °C. Transfer the clean supernatant (crude chromatin) to a new tube (using siliconized tubes significantly reduces carryover of non-antibody bound chromatin). (*see* **Note 11**.)

3.3 Immunoprecipitation

1. Prepare the buffer for immunoprecipitation by adding one tablet of PhosSTOP to 10 mL of IP buffer.

2. Mix the solubilized chromatin with an equal volume of the buffer from Subheading 3.3, **step 1** (750 μL of each). Set aside 120 μL of solubilized chromatin to serve as "input DNA" (store at −20 °C). At this point, the supernatant can be divided in two equal volumes to be used as technical replicates of the immunoprecipitation.

3. Incubate the protein–DNA complexes with 2 μg antibody recognizing one of the Pol II forms for 4 h to overnight at 4 °C on a rotation wheel (*see* **Notes 12** and **13**).

4. Centrifuge at top speed for 5 min at 4 °C to get rid of debris (*see* **Note 10**).

5. Transfer the supernatant to a new 2 mL DNA LoBind tube. Capture the immune complexes by adding 30 μL of protein-A magnetic beads (previously equilibrated in ChIP IP buffer) and incubate at 4 °C for 1.5–2 h on a rotation wheel.

6. Separate the beads using a magnetic stand. Discard the supernatant without disturbing the pellet (antibody–protein–DNA complexes bound to beads) and continue with the beads.

The supernatant can keep at –20 °C and reprocessed if the immunoprecipitation does not work.

3.4 ChIP Wash and Chromatin Elution

1. Wash the beads five times with 1 mL of IP buffer for 10 min on a rotation wheel (the first two washes at 4 °C and the rest of washes at room temperature). Recover the beads between the washes using a magnetic stand and remove the solution. Be sure to pipette off all liquid in the last wash (*see* **Note 14**).

2. Elute the protein–DNA complexes from the beads by adding 100 µL cold elution buffer, vortex vigorously, incubate for 1 min at 37 °C and pellet the beads with a magnetic stand. Transfer the supernatant (eluted) to 1.5 mL microcentrifuge tube. Add 50 µL of 1 M Tris–HCl, pH 9, to neutralize. Repeat the elution and neutralization twice more, with the last elution step being performed for 4 min at 37 °C. The combined volume of eluate will be 450 µL.

3.5 Reverse Crosslinking and DNA Purification

1. Add 1 µL of RNase A/T1 and incubate for 15–30 min at 37 °C. Then add proteinase K to final concentration of 0.5 mg/mL and incubate overnight at 37 °C. (This should be done with "input DNA" from Subheading 3.3, **step 2** as well as with the immune precipitates. For this, the "input DNA" should be brought to a final volume of 450 µL with TE buffer.)

2. Add a second aliquot of proteinase K (same amount as in Subheading 3.5, **step 3**) and incubate at 65 °C for at least 6 h to reverse the formaldehyde crosslinks between proteins and DNA.

3. Purify DNA using MinElute PCR purification kit, following the directions to the letter. Elute the DNA into low-adhesion tubes in 30 µL of elution buffer. Purified ChIP sample can be stored (*see* **Notes 15** and **16**).

4. Carry out quantitative PCR to test ChIP enrichment (*see* **Note 17** for comments regarding choice of gene and primer design). Dilute the eluted DNA from Subheading 3.5, **step 3** 1:5 with Milli-Q water, add 5 µL of diluted DNA to 5 µL of primer master mix (1 µM per primer, final concentration; *see* **Note 17** for comments about primer choices), 12.5 µL of iQ SYBR Green Supermix, and water to bring the final volume to 25 µL. Execute the reactions using cycling conditions predetermined to work with the primer combinations (*see* **Note 17**).

5. Calculate the relative enrichment for each amplicon using the comparative Ct method [13], where qPCR data is normalized using input DNA and corrected for the DNA that purifies with immunopurified IgG. In this way, it will be possible to have a quantitative estimate of Pol II distribution across of different candidate genes.

3.6 Next Generation Sequencing

Sequencing libraries may be prepared using any of a number of commercial kits for the production of libraries for high throughput sequencing. Typically, library construction entails a series of steps that result in the addition of platform-specific sequences to the ends of short DNA fragments, such as are generated by the ChIP protocol described in Subheadings 3.1–3.5. There is usually an amplification step, after which additional measures may be taken to improve the yields and qualities of libraries. As an example, the following describes library preparation using one commercial kit, along with subsequent steps.

1. Use 5–25 ng of immunoprecipitated DNA to prepare the library. Run an aliquot of the ChIP DNA on an Agilent Technologies 2100 Bioanalyzer using a high sensitivity DNA chip to verify the size and concentration of the DNA. If necessary, material from several immunoprecipitations may be pooled to obtain the required quantity (at least 5 ng) of starting DNA.

2. When beginning with 5–10 ng of starting immunoprecipitated DNA, libraries are made using the NEBNext® Ultra™ DNA Library Prep Kit for Illumina. When beginning with greater than 10 ng of immunoprecipitated DNA, theNEBNext® ChIP-Seq Library Prep Reagent Set for Illumina is used. The adapters (*see* **Note 18**) and primers (*see* **Note 19**) necessary during this protocol can be found in NEBNext Singleplex or Multiplex Oligos for Illumina. The instructions provided by the supplier are followed; prior to PCR amplification of the library (performed according to the supplier's recommendations), the adapted DNA fragments are purified using AMPure XP beads as recommended by the supplier of the library preparation kit.

3. Separate the PCR-amplified library on a 2 % agarose gel at 120 V for 10 min, then 60 V for 180 min. Photograph the gel and then excise the region corresponding to 200–500 bp (*see* **Note 20**). Photograph the gel after excision to document to excised region. Place the gel slice in a new DNA LoBind tube.

4. Extract the DNA from the gel using the MinElute gel extraction kit following the instructions. Use 6 volumes of Buffer QC per volume of gel and incubate the gel slices at room temperature with periodic vortexing until the gel slices are dissolved (*see* **Note 21**). Add 2 volumes of isopropanol, and follow the kit instructions. Elute in 25 µL of buffer EB.

5. Amplify the library by PCR using the following conditions: 30 s initial melting at 98 °C; 12–20 cycles (*see* **Note 22**) of 10 s at 98 °C, 30 s at 65 °C, and 30 s at 72 °C; and a final extension step of 5 min at 72 °C. For these reactions, use 16 µL of the eluted DNA from Subheading 3.6, **step 4** as template, 10 µL 5× Phusion HF buffer, 50 pmol of each primer

(added from 100 µM stocks), 1.5 µL of 10 mM dNTP mix, 0.5 µL of Phusion DNA polymerase; the final volume is adjusted to 50 µL with DNase/RNase free water.

6. Add 50 µL of AMPure XP beads and mix thoroughly. Incubate at room temperature for 8 min. Collect the beads using a magnetic stand, and wash the beads twice with 200 µL fresh 80 % ethanol. Elute the DNA in 20–30 µL of buffer EB (use a 1.5 mL low-binding tube). This is the sample to be submitted for sequencing.

7. The final eluate may be assessed in two ways. One is to test for specified target enrichment using qPCR as in Subheading 3.5, **steps 4** and **5**. Additionally, DNA size distribution, purity and concentration may be determined using a Bioanalyzer with a high sensitive DNA chip. Libraries that show reliable target enrichment and that have the expected size distribution (without low or high molecular weight artifacts; *see* Fig. 1) are suitable for sequencing.

4 Notes

1. This protocol has been used with *Arabidopsis thaliana* (ecotype Columbia) plants grown in normal conditions (Murashige and Skoog (MS) salt, supplemented with 10 g/L sucrose, 0.5 g/L MES and 8 g/L agar, pH 5.6–5.7 during 12 days at 20/18 °C under 16 h light–8 h dark regimen), but in principle, it can be used for virtually any tissue developmental stage or growth condition or treatment.

2. The genome-wide distribution of Pol II is determined using four antibodies (*see* **Note 11**). One immunoprecipitates all forms of Pol II (anti-RNA polymerase II CTD repeat YSPTSPS antibody), the others are specific for the phosphorylation of the Pol II at ser5 (anti-RNA polymerase II CTD repeat YSPTSPS, phosphor S5), ser2 (anti-RNA polymerase II CTD repeat YSPTSPS, phosphor S2) or ser7 (anti-RNA polymerase II CTD repeat YSPTSPS, phosphor S7). Rabbit control IgG antibody is used as negative control.

3. One complete ULTRA Tablet, EDTA-free is sufficient for a volume of 50 mL extraction solution and one complete MINI tablet EDTA-free is sufficient for a volume of 10 mL extraction solution.

4. The tissue collection should take no longer than 1 h to prevent withering. Stability of the plant material is dependent on the tissue, e.g., on its size and water content.

5. During the fixation step, to improve the process, release vacuum at 15 min intervals, mix the tissue by inverting the tube

several times, and reapply vacuum infiltration. Repeat this process three times, so that the total fixation time is 1 h. During this process, be sure that the tissue is submerged in the buffer and not floating on the surface.

6. At this point, the tissue can be stored at −80 °C or powdered under liquid nitrogen to proceed with the nuclei isolation.

7. Prior to use, humidify the meshes with M1 buffer. Use 1 mesh per sample. Be sure to squeeze the Miracloth after the last pass-through to extract all of the liquid. Always keep the samples on ice.

8. An efficient sonication is crucial for a good enrichment and resolution in ChIP experiments. The resolution of ChIP is ultimately limited by the average fragment shearing size (an average fragment size of 200–800 bp is recommended). The target shearing size must be determined empirically for each chromatin preparation. The degree of crosslinking and viscosity of the chromatin solution influences the efficiency by which the ultrasonic waves shear the chromatin [14].

9. Time of sonication depends on sonicator and time of fixation. With a Bioruptor UCD-200 (Diagenode), the sample is sonicated with the pulse on high for 30 s, followed by 30 s of rest; this is repeated for a duration of 15 min. The water bath should be maintained at a constant temperature of 4 °C. Alternatively, using a probe sonicator (Fisher, Model 300 sonic dismembrator), the sample is treated for 4×15 s pulses at half maximal power with extensive cooling between pulses, making sure that the sonicator probe tip does not touch the side of the tube.

10. Do not transfer any insoluble material to new tubes after centrifugation to avoid low enrichment in the ChIP-DNA. If necessary, repeat this step.

11. Before proceeding to the immunoprecipitation, the shearing size of the extract should be confirmed by reversing the crosslinks (*see* Subheading 3.5), purifying a small aliquot of the extract (as in Subheading 3.5), and analyzing the DNA on a 1.5 % agarose gel.

12. The specificity and sensitivity of the target factor enrichment is a key determinant of a successful ChIP experiment. Specific protein–DNA intermediates are enriched by immunoprecipitation using antibodies against the native protein. The antibodies must be specific for the protein of interest as well as recognize the native form of the protein when associated with chromatin. Use high quality antibodies are advisable and critical for ChIP assays. In our hands, polyclonal antibodies from Abcam give very good results (specifically targeted against one of the different un/phosphorylated forms of RNA polymerase II CTD), and usually give higher ChIP signals than a mono-

clonal antibody. A nonspecific antibody as well as IgG as negative controls may be used to discriminate true binding sites from regions that may interact promiscuously with the antibody or beads in a nonspecific manner.

13. The immunoprecipitation step can be left overnight and is thus a good stopping point for the end of the day.

14. The IP buffer should be removed carefully from the magnetic beads without disintegrating the bead pellet. Remove as much of the IP buffer as possible in the washes to avoid low enrichment in the ChIP-DNA. From this point, the samples can be handled at room temperature.

15. Since the quantity of DNA recovered after ChIP is characteristically very small, the results of any downstream analysis are extremely sensitive to sample degradation, and immunoprecipitated DNA should be analyzed as soon as possible. If not amplified immediately, the DNA should be stored at –20 °C, but for no more than 1 month. Multiple freeze-thawing should be avoided.

16. The number of times that a ChIP experiment needs to be repeated depends on the fold-enrichment achieved and experimental variance. The variance of an experiment is specific to each experiment, and is difficult to model and generalize. In most cases, the aim of repeating an experiment is to determine which part of the signal has a biological meaning. Typically, three fully independent ChIP experiments are performed using samples collected separately.

17. There are several considerations when designing qPCR experiments, including selection of an appropriate set of reference regions, exact primer localization, and methods for quantification of the results. Routinely, a target gene that is median in terms of length (long enough to permit distinction of different genomic regions) and expressed at relatively high levels (to facilitate the PCR reactions) is chosen. Optimally, primer pairs should be designed to amplify 60–100 bp products within the central portion of the identified regions. Test primers on genomic DNA for efficient amplification prior using them on ChIP reactions.

18. An excess of adapters and primers leads to the potential to form dimers that may compete with the final library amplification, resulting in a 100 bp band that may contaminate the sequencing reaction. For this reason, primers are used at a final concentration of 1.5 μM [13].

19. It is possible to multiplex or "barcode" multiple samples [14] in a single sequencing lane, and thus to sequence multiple samples in a single lane.

20. Clean the tray, the comb and the gel tank with ethanol and rinse thoroughly with deionized water to avoid cross-contamination. When the sample is loaded, leave a gap of at least one empty lane between samples and ladders. If the sample prep and amplification are successful, when the gel is visualized, a smear of DNA will be readily apparent.

21. Dissolving the gel at 50 °C, as recommended in the kit protocol, may result in denaturation of the library and a loss of AT rich sequences; therefore, it is recommended that the gel is dissolved at room temperature.

22. The optimal number of cycles depends on the amount of DNA in the original ChIP sample and can be varied between 12 and 20 cycles. Typically, 18 cycles will give good results when starting with 10 ng of input DNA. Typical results are shown in Fig. 1.

Acknowledgements

The author is supported by NSF grant MCB-1243849. The author thanks Prof. Javier Paz-Ares at National Centre of Biotechnology, Madrid, Spain for his helpful comments on this manuscript.

References

1. Lunde BM, Reichow SL, Kim M, Suh H, Leeper TC, Yang F, Mutschler H, Buratowski S, Meinhart A, Varani G (2010) Cooperative interaction of transcription termination factors with the RNA polymerase II C-terminal domain. Nat Struct Mol Biol 17(10):1195–1201. doi:10.1038/nsmb.1893

2. Buratowski S (2009) Progression through the RNA polymerase II CTD cycle. Mol Cell 36(4):541–546. doi:10.1016/j.molcel.2009.10.019

3. Egloff S, Murphy S (2008) Cracking the RNA polymerase II CTD code. Trends Genet 24(6):280–288. doi:10.1016/j.tig.2008.03.008

4. Hsin JP, Manley JL (2012) The RNA polymerase II CTD coordinates transcription and RNA processing. Genes Dev 26(19):2119–2137. doi:10.1101/gad.200303.112

5. Egloff S (2012) Role of Ser7 phosphorylation of the CTD during transcription of snRNA genes. RNA Biol 9(8):1033–1038. doi:10.4161/rna.21166

6. Kim M, Suh H, Cho EJ, Buratowski S (2009) Phosphorylation of the yeast Rpb1 C-terminal domain at serines 2, 5, and 7. J Biol Chem 284(39):26421–26426. doi:10.1074/jbc.M109.028993

7. Phatnani HP, Greenleaf AL (2006) Phosphorylation and functions of the RNA polymerase II CTD. Genes Dev 20(21):2922–2936. doi:10.1101/gad.1477006

8. Johnson DS, Mortazavi A, Myers RM, Wold B (2007) Genome-wide mapping of in vivo protein-DNA interactions. Science 316(5830):1497–1502. doi:10.1126/science.1141319

9. Park PJ (2009) ChIP-seq: advantages and challenges of a maturing technology. Nat Rev Genet 10(10):669–680. doi:10.1038/nrg2641

10. Lubelsky Y, MacAlpine HK, MacAlpine DM (2012) Genome-wide localization of replication factors. Methods 57(2):187–195. doi:10.1016/j.ymeth.2012.03.022

11. Saleh A, Alvarez-Venegas R, Avramova Z (2008) An efficient chromatin immunoprecipitation (ChIP) protocol for studying histone modifications in Arabidopsis plants. Nat Protoc 3(6):1018–1025. doi:10.1038/nprot.2008.66

12. Yamaguchi N, Winter CM, Wu MF, Kwon CS, William DA, Wagner D (2014) PROTOCOLS: chromatin immunoprecipitation from arabidopsis tissues. The Arabidopsis book. Am Soc Plant Biol 12:e0170. doi:10.1199/tab.0170

13. Livak KJ, Schmittgen TD (2001) Analysis of relative gene expression data using real-time quantitative PCR and the 2(-Delta Delta C(T)) Method. Methods 25(4):402–408. doi:10.1006/meth.2001.1262

14. Auerbach RK, Euskirchen G, Rozowsky J, Lamarre-Vincent N, Moqtaderi Z, Lefrancois P, Struhl K, Gerstein M, Snyder M (2009) Mapping accessible chromatin regions using Sono-Seq. Proc Natl Acad Sci U S A 106(35):14926–14931. doi:10.1073/pnas.0905443106

INDEX

Arthur G. Hunt and Qingshun Quinn Li (eds.), *Polyadenylation in Plants: Methods and Protocols*, Methods in Molecular Biology,
vol. 1255, DOI 10.1007/978-1-4939-2175-1, © Springer Science+Business Media New York 2015

"

If you have any concerns about our products,
you can contact us on
ProductSafety@springernature.com

In case Publisher is established outside the EU,
the EU authorized representative is:
**Springer Nature Customer Service Center GmbH
Europaplatz 3, 69115 Heidelberg, Germany**

Printed by Libri Plureos GmbH
in Hamburg, Germany